I0484666

ON THE WAY OF
THE GREAT SCIENTISTS

ON THE WAY OF
THE GREAT SCIENTISTS

A Simple Guide For The Science Way

By
Sherif M. Elkaffas

2015

Copyright © 2015 by Sherif Elkaffas

All rights reserved.

This book or any portion thereof may not be reproduced or used in any manner whatsoever without the express written permission of the publisher except for the use of brief quotations in a book review or scholarly journal.

ISBN: 1507505272

ISBN-13: 978-1507505274

Email:sherif.mostafa87@yahoo.com

THIS BOOK IS DEDICATED TO MY PARENTS,

without their continuous support and sacrifice, this book couldn't have been written

"Remember our servants Abraham, Isaac, and Jacob possessors of strength and vision."

Quran (38:45)

Table of Contents

Get Out Of Your Prison

Most of us are prisoners of our minds. We built these prisons by our negative thoughts and negative expectations about ourselves and our destiny. All the time we think about failing and absence of abilities, skills or qualifications. We always doubt about what is possible. We are staying there in our prisons, without moving and without using the gifts that God has given us. Preventing the world from seeing what we can offer for it. While we are there, other individuals didn't surrender and destroyed all the limitations of self-doubt. They believed in themselves, even if others were not. They determined to complete their way, even if there would be obstacles and difficulties. The great scientists belong to those individuals. They overcame all the difficulties and believed in themselves, after that they got the ability to change the world. Also, you can overcome all self-doubts and get out of your prison. The first step is to learn and know more because knowledge is the freedom gate.

Preface

During my early years of study I was wondering about the scientific discoveries which were taught to me. After every new lesson about a scientific discovery made by a scientist, I was asking myself who are scientists? How they made these different amazing and wonderful discoveries. I was asking, are they a special kind of humans with special capabilities of thinking? I was interested to know more about them and to read their life stories and the stories behind their discoveries. I wanted to know how they begin their journey in science and how they are thinking and how great ideas are formed. After reading a large number of the great scientists' life stories, searching for the answers of my questions, I found that there is a common way that all distinguished scientists walked through it. This way consists of different steps. In this book, I tried to clarify these different steps for helping who is interested in science to know the secrets of the great scientists.

Our universe is filled with an infinitive number of secrets. The scientists are individuals who are trying to discover these secrets by doing scientific researches. Today, we are surrounded by amazing inventions that are based on the discoveries and the works of the great scientists. This book is a simple guide to know how the great scientists made these wonderful and amazing discoveries. It includes stories of wonderful life situations of famous scientists and also includes advices and successful strategies extracted from their life stories. It can help who is working in scientific research to find his way to make wonderful discoveries. **It's no longer a secret, now you can know how they discovered these wonderful discoveries and also you can.**

1

Step One:
Follow Your Passion
for Science

Passion for science is the most obvious characteristic of the great scientists. Not only that they have a great passion for science, but they were also brave enough to follow it.

1.1. The Way of Success....

"Follow your passion and success will follow you"

Arthur Buddhold

All the time, we hear that if we follow our passion we will succeed. This is the most common advice that you can get from the successful people. Also, when you look around searching for the most successful people you will find them working in what they love and following their passion. So, that's a true. There must be a strong relation between passion and success. But what are the secrets behind this strong relation?

This is really an important question, first we have to know what the success is. Success in life is a desire of everyone. We all search for happiness and we think that success is a great source for happiness, so we all want to succeed in what we do. We search for it in all different life disciplines: in work, family, etc. But, it is very difficult to define what success is because everyone has his own meaning of success. In general, success could be defined as the achievement of the desirable results. To succeed in what you do is to achieve the goals that you have set for yourself and it is not easy to get them. To succeed you need to begin with a high degree of preparation, then hard working with some degree of creativity and the most important thing is to be very motivated to get what you want. Here, the role of passion appears because the most important thing that can help you to succeed in your work and to not give up is to have a passion for what you do.

I want to tell you an advice from someone that I think many people in this world were inspired by him, he is Steve jobs. When Steve Jobs was asked about the relation between success and passion he mentioned that success needs many works and only passion about what you do will help you to not give up. So, everything you want to succeed in required you to have passion about it. You need to have a passion about what you do, either about science or any other discipline, in order to succeed in it.

What Can Passion Do?

Your passion about what you do and your love for it will increase your chances for success. But what passion can do? The reasons behind this strong relation between passion and success appear in the different roles of passion.

Passion and the costs

First, passion will help you to bear the costs, either time or effort, of the work. There are different costs that you must pay to succeed. No success without spending great efforts or without hard working. This hard working needs physical and mental efforts plus a need for spending time. If you have a passion about what you do, it will help you to sustain and bear these costs. As it was said if you find a work in what you love, you don't need to work for the rest of your life. Who have a passion about what he do will easily spend physical and mental efforts on it because he love this work and will not feel any tired instead he will enjoy with his work. Who work in what he loves will spend more efforts and time easily in contrast to others who don't find passion about it and this is a big advantage. Many distinguished scientists refuse the breaks and relaxation and they were spending a lot of time in their work. Thomas Edison is a clear example of that. Nicolas Tesla worked for a period as an assistant to Thomas Edison and he described the behaviour of Edison during this period and said; *"I came from Paris in the spring of 1884, and was brought in intimate contact with him (Thomas Edison). We experimented day and night, holidays not excepted. His existence was made up of alternate periods of work and sleep in the laboratory. He had no hobby, cared for no sport or amusement of any kind and lived in utter disregard of the most elementary rules of hygiene. There can be no doubt that, if he had not married later a woman of exceptional intelligence, who made it the one object of her life to preserve him, he would have died many years ago from consequences of sheer neglect. So great and uncontrollable was his passion for work."*

Passion and the other attractions

Second, passion will help you to resist other attractions. If you have a passion about something, it will be very hard to leave it for anything else, even if the other thing has more substantial or financial profits. That could be clarified by master Alakad who was one of the most famous and successful Arabic writers. He described this case, when some people said to him to leave the field that he was passion about and work in another field that can help him to get more money. He clarified that the writer who leaves his passion for literature and choose to work in another field to get more money is like a father who replaces his bad son with other foreign good boy. Here, he described his relation to what he was passion about as the relation between a father and his son. As the father can't accept the replacement of his son with any other boy even if he is much better, also who have a passion don't accept to leave it for any other thing. So, it's a big advantage if you have a passion about what you do because without it, it will be so hard to resist the other attractions.

Passion and the creativity

Third, passion will not only help you to continue on your way but will also help you to be more creative in your field than others who don't have passion in this field. This is the last and the most important advantage about following your passion. This increase in creativity is due to the effect of Dopamine. Alice Flanerty, a neuroscientist interested in studying creativity, said "People vary in terms of their level of creative drive according to the activity of the dopamine pathways of the limbic system". So, Dopamine has a great influence on increasing the creativity. Recent studies showed that this dopamine is related to our feelings and its level increased when we are happy or relaxed or feels good. So, the reason behind the increasing in the creativity is the happiness that we feel when we work in what we are passion about. This happiness stimulates the high level of Dopamine. Then this dopamine is taken up by certain brain areas and makes these areas more active and increases our creativity. This effect also explained by professor Richard Depue the director of the laboratory of neurobiology of personality and emotion at Cornell university "When our dopamine system is activated we are more positive, excited and eager to go after goals or rewards, such as food, sex, money, education or professional achievements." Also, this effect of dopamine explains why we get creative idea when we have a good feeling.

In The End

Passion about what you do, either science or any other field, increases your chances of success. It will help you in working hard easily and to not give up for any other temptations and also will increase your creativity. And this is may be the reasons that explain why individuals succeed in what they love and what they have passion about.

1.2. Passion Is The Start Point....

"When I hear a new word of science that I didn't hear before, I wish that all my body organs become ears to share the wonderful feelings that my ears get with this new knowledge."   Alshafi'i

As you have known that most of the successful people work in what they love and in what they are passion about, it's also the same for the great scientists that passion is the secret behind their success. I am very interested in reading the life stories of the distinguished scientists. I read a large number of their stories and I found that the most common recognized characteristic between them is their passion for science.

"Passion for science", these are the words that we use to express the highest degree of love knowledge and learning. Passion for science is the real reason behind the great scientists' choice of the science way. I think that the presence of this passion is what differentiate between who are a real scientist and who are not. Many people may choose to work in science and scientific research due to different reasons such as searching for scientists' prestige and the respect that scientists have or searching for money or to be famous. But distinguished scientists choose to work in scientific field only because of their passion for science. It represents the start point of their way in science.

Examples for The Passion for Science

If you read a life story of anyone of the great scientists you will see the clear presence of this passion for science. These are few examples that express the presence of this passion for science.

The early beginning of Faraday

Faraday, who was one of the most distinguished scientists, was characterized by a great passion for science. Faraday's passion for science was begun at his age of fourteen. Before his joining to the lab of Sir Humphrey Davy, Faraday was working in the bookbinding trade. During this period, he was interested to learn and teach himself by reading nearly all the books that he could reach during his work as a bookbinder. He wrote about these early days: *"Whilst an apprentice I loved to read the scientific books which were under my hand."* One of these books was Mrs. Marcet's Conversations in Chemistry which excited his interest. He also read Watts' On the Improvement of the Mind which made him think. Then he read an article on Electricity in the Encyclopedia Britannica and this book is considered as his start to science.

Newton and the farm

Isaac Newton was also characterized by this early beginning of passion for science. Isaac Newton was prepared to be a farmer to help his mother Mrs. Newton in her work. When his mother sent him to watch sheep in the field, he was spending his time staying under a tree reading a book or making some models. This behavior of Isaac Newton of loving learning and knowledge was also clear, when his mother was sending him with a good servant to Grantham on each market day to do some works for the farm. But when Isaac reached Grantham, he was leaving the old servant to look after the marketing and went to the Apothecary's house, to spend his time among the chemist's stock of books. He was still there until the call for the return journey. This behaviour made Mrs. Newton consult her brother about the future of Isaac and his uncle suggested that Isaac should return to Grantham to be prepared for college and the idea of farming is not appropriate for him.

Aljahez reading nights

Another interesting behavior about the high degree of loving learning and knowledge was the passion of Aljahez for acquiring knowledge. Aljahez was one of the most brilliant Muslim minds in the golden age of Islamic civilization. He was very keen to read all the books of his time, but he didn't have enough money to buy all the books that he wanted. For that, he was renting the bookshops to sleep in at night and spending his time in reading the books that he couldn't buy.

The great enjoyment

You can see the passion for science among nearly all the great scientists, regardless their specialization fields. They found a great enjoyment in learning and in acquiring new knowledge that made them highly passion for science and scientific research. Marie Curie described her passion for science by saying: *"I am among those who think that science has great beauty. A scientist in his laboratory is not only a technician: he is also a child placed before natural phenomena which impress him like a fairy tale"*. Marie Curie described the scientist in his lab as a child. Some scientists tried to describe their wonderful feeling about science like Pasteur who said; *"Science…it is my life…it has brought me a deepness of pleasure that I have always known yet never realized."* Ibn Alkiem also said; *"If science has a picture, it will be more beautiful than the sun and the moon"*

In The End

From all the previous examples, now I think it's clear to you why scientists choose to work in the scientific field and to enter this way. Although there are many advantages that many people may find about being a scientist, the distinguished scientists' main reason is their passion for science. They choose the way of science because they love to work in it, not searching for anything else.

1.3. Passion Beginning....

"All men by nature desire to know."
Aristotle

After you have known that the passion for science is the most commonly recognized character between nearly all the distinguished scientists and how passion can help in increasing the chances of success, there are important questions; what is the source of the passion for science?. What are the reasons behind it? How is it formed? How do the great scientists have this high degree of passion for science?

The answer is unexpected. Surprisingly, all of us have to some degree a passion for science. Yes, that's a true, you and me and all humans have it. It's born with us since our birth moment. It's responsible for making all humans want to know, want to learn, want to discover and want to understand all things in the universe. This is the curiosity that all people have. This curiosity begins from our childhood. If you look at any child you will find him trying to discover anything that his hand can touch. All the time, he watched the movements of anything around. Also, when we were children, this curiosity made a lot of troubles for us.

The Developing of the passion for science

The curiosity is scientifically defined as an intense, intrinsically motivated appetite for information. William James divided it into two main types:

- Common curiosity: this type includes the excited feelings brought on by the novelty.

- Scientific curiosity: this is the type related to exciting feelings about a specific type of information.

Although we all born with the common curiosity, the scientific curiosity needed to be stimulated. This scientific curiosity is the motivational factor for the development of our passion for science. This scientific curiosity could be increased or decreased. In the great scientists, this scientific curiosity reached high levels more than normal people. This Increase or decrease of curiosity depends on many factors which mainly begin during their early stages.

One of the main factors that stimulated and enhanced this scientific curiosity in many of the great scientists was the role of their family especially the father. It could be explained from two of the distinguished scientists. First one is Einstein and his father compass. When Einstein was 5 years old and while he was sick in the bed his father gave him a compass. He was amazed by it and wondered how the needle of this compass always pointed toward the north. This compass stimulated his scientific curiosity. This compass made him very interested about the invisible powers that work behind the objects. The second example about the role of the father in stimulating the scientific curiosity is Louis Pasteur. His father, Jean Pasteur was a tanner without much education. Louis Pasteur clarified the role of his father in stimulating his passion for science, in one of his letters to his father. Pasteur wrote: *"You (Pasteur's father) might not remember how important your influence on me was in developing my mind... It was you who helped me decide to study natural sciences—undoubtedly because of your own interest in the subject rather than your conviction regarding my aptitude. Enthusiasm and mother's presence of mind were all passed on to me by you. If I have always associated the grandeur of our country it is because of the feelings that you inspired in me."*

Also, teachers have a very important role in increasing the scientific curiosity of the students, especially in their early learning stages. The way and method of teaching could help in increasing or decreasing this curiosity.

In The End

Although the scientific curiosity of the great scientists was stimulated during their early stages, this curiosity could be stimulated at any stage. Reading widely in science, attending the public scientific lectures or watching the scientific documentary films could stimulate and increase this scientific curiosity.

1.4. Take The Brave Decision....

"You are a possibility that has never occurred before and will never occur again. No one else has had or will ever have your unique combination of talents, experiences and dreams. So don't waste that uniqueness."

Patrick Combs

Now, you know the importance of the passion for science in achieving what the great scientists have achieved. But they not only had a great passion for science, they were also brave enough to follow their passion. So, if you have a passion for science you have to follow it. I talked before about the strong relation between passion and success. Now, you know that following your passion will help you to increase your chances of success. But it's not the only reason to follow your passion for science. There is another important one. It is your responsibilities that came from your unique abilities. You are a special individual with special abilities. Yes, you are. It's a fact; everyone is a special person in his experiences, talents, skills and goals. Do you think that anybody else has your unique collection of abilities? Actually, no one has. And by these abilities you can help many individual live a better life. This specialty is not only an advantage for you to others, but it also has a responsibility that you must take. Taking these responsibilities is the most important reason to follow your passion for science. **The worst thing in life is a person who was able to change the life of many individuals to be better and didn't do.** Look at your talents, your skills and your experiences what will you do with them.

Your Positive Imprint

There are two types of individuals in this life; first type is the individuals that only think about themselves **"The Normal People"** and the second one is the individuals that can think of others **"The Great People"**. The great people always have a positive imprint in the life of others. They are the individuals who have a purpose for their life. They have something valuable that they try to do and the most valuable thing is to help other people to live a better life. Also, you have to search for what valuable things that you can do.

Change The World

You must know that everyone in this life has a role and this role is his life mission. This role is mainly based on his abilities that he acquired during his different life stages. So, you must take your responsibility and do your role, according to the abilities that you have. Your decision to follow your passion for science may be the first step in changing the world to be better. Many great scientists began their way to amazing and wonderful discoveries by taking this decision. Then they discover things that change the way that we look at the universe and they are remembered for hundreds of years.

Your Passion Is A Signal

The truth about the specialty of the great scientists is that; **"They are not smarter than you, but they are brave enough to take the decision to follow their passion for science."** Following your passion needs a brave decision. That's a true; this step needs a brave person. I want to ask you a question, have you ever met before someone who was very smart, talented and has unique abilities but he didn't use it in any useful thing? I am sure that you have met many persons like that. But do you know why they became like that? The secret is their fear to leave their safe area. Also, the reason that made you stay in your place and prevent you from doing your role in life and from following your passion is the fear. Yes, you are afraid to follow your passion. Do you know why? Because, you don't want to take the risks or to leave your safe area. You may now have a normal life like working in a job only for money with a regular salary every month. You nearly don't take any risks in your life. What you normally do is to do the tasks of your job and to get your salary. You are in a safe area **but it's not your place.** Actually, all the changes that they can make your life better is outside your safe area. If you still in your safe area you will live a life less than the real life that you worth. Also, you are preventing other people from the benefits that can get if you follow your passion for science.

You may have fears about the costs of the risks that you should take and you don't know if you will succeed or not. But, there is a law that you must know and be sure of it that **"If you have a passion for science and moved toward it, you should trust that god will help you to succeed."** This feeling which we call passion for science is the signal that god give us to know our place and our role in this life. I can't promise you that success will be easy, but I can promise that if you persist you will get it.

In The End

Before taking your decision for following your passion for science or not, you have to consider the following points. First, many people took the decision to follow their passion for science, then they changed the life of many other individuals to be better. Also, your abilities can change your life and others life to be better, if you follow your passion for science. Second, your safe area is not safe anymore. Third, trust in that your passion is your guide to know your life mission. Finally, no pain no gain, there must be a cost to follow your passion, but believe me **the costs of not following your passion will be higher to you and to mankind.**

2

Step two:
Choose Your Field

Your time and effort are very limited. They are not enough for you to acquire all the available knowledge. So, you have to choose a specific field of study.

2.1. Specialization Importance....

"If you chase two rabbits, both will escape"

Unknown author

One day, I read a story that completely changed my mind. Before reading it, I was believing like many people that the more talents I have, the more chances of success I can get, but I found it is a wrong idea. This story was mentioned by Paulo Coelho in one of his interesting books. It is a story about a prince and three fairies that were invited to his baptism. Each fairy gave the prince a gift. They tried to make him happy in his life. First one gave him the ability to find his love, second one gave him enough money to do as he pleased and the last one gave him beauty. But, there was a witch one that determined to corrupt all these gifts. She gave him the ability to be talented at whatever he tries to do. The prince grew up with all the talents that he wanted and he was an excellent painter, sculptor, musician, mathematician -- but he couldn't complete a task because he quickly became distracted and wanted to move on to something else. This gift alone was enough to fail him and made him lived a very sad life. The story was finished with an important quote that: *"All roads lead to the same place. But choose your own, and follow it to the end. Do not try to walk every road."*

From this moment, I corrected my thoughts and recognized that success, in any life disciplines, needs specialization. You and I like all humans, we have limited time and effort and absence of focusing will lead to failure. If you involved in different directions or activities you will not be able to give any of them the appropriate amount of concentration and you will not succeed in any of them. Also, Success in science needs specialization.

Choose One Narrow Area

In the past, it was common that scientists might be involved in different scientific areas and you might hear about many scientists who made studies and wrote books about philosophy, chemistry, medicine and astronomy at the same time. But now it's impossible to be specialized in all these fields because the amount of the available knowledge in the past was not as the available knowledge today. Nowadays science is very wide and you cannot acquire all the available knowledge. So, you must get specialized in one narrow area of Science.

Wrong Meaning of Specialization

You should take care because there is a wrong meaning that many can get when they hear about specialization. Some people think that specialization in science is to focus only on your narrow area without knowing about other disciplines. That's a wrong understand. Nowadays, there are high interactions between different fields. For example, to study biology you actually study phenomena based on chemical or physical facts. It's essential and not optional to be aware of the basics of other scientific branches and so on in other fields. So, the researchers not only should know the general knowledge about their wide areas like chemistry and its different branches that deals with it, but also they need to have knowledge about other basic field of science to be able to understand their narrow area of science.

Most of current scientific research teams have different specialization. They may have one for statistics and another for IT and other one in any related field to cooperate together in order to solve the scientific problem. So, specialization here is to know your field and to have some basic knowledge about other fields. This knowledge could help in finding important discoveries that couldn't be done without cooperation between these different disciplines.

Every researcher can only work in a narrow area, but they should have a minimal amount of knowledge about other disciplines, especially the basic scientific fields which is (Mathematics, chemistry, physics, and biology). Because today the great breaks through are made by a physical researcher who help in enhancing or discovering something useful in biology and a biological researcher who help in understanding a chemical problem and mathematician who explain a physical phenomenon and so on.

Human knowledge is divided into different wide areas, each wide area has different branches within it and each branch contains narrow areas. So, you need to know the basics of your wide area and what the narrow areas within them are and the details about your narrow area and its related fields. Also, you need to know the general basics of other areas of science which means to know (What's the goal of it? - What's it deal with? – What are the fields of study in this area?). You can based on this information not only relate any information to its area, but also you can relate it to its specific field within this area. And when you need to know a part of another field in order to clearly understand a specific point in your area, you will easily know what the specialization of the person who can help you in this point.

In The End

To succeed in science you should be specialized in a narrow scientific area and have a good level of knowledge about other scientific areas especially the basic and the related ones.

2.2. Find Your Interest....

"There are only two forces that unite men - fear and interest".

Napoleon Bonaparte

The choice of your scientific area is a very important decision and should be taken very carefully. You should be highly interested in it because your scientific area will be the work that you will spend the rest of your life in. Steve jobs said *"your work is going to fill a large part of your life and the only way to be truly satisfied is to do what you believe is great work and the only way to do great work is to love what you do."* Finding this interest is the motivational key that will help you to continue and to success in your way.

What's the interest?

Many people have a problem in finding their interest and they don't know how to find it. It is really a very important question, but before answering this question there is another important question that should be answered first that "What is the meaning of interest?" what do we mean by saying that we are interested in something? Answering this question is the start point that will help us to find our interest.

We say that we are interested in something when we want to express our good feelings about it. These good feelings made us care about this issue and want to be involved in it. These good feelings are formed due to the presence of something that touched our hearts and attracted our attention which we can call it a hook. So, the hooks are the main reasons for loving what we love.

The hooks of Interest

There are different types of hooks that could affect our choice of a specific scientific area and every hook has different degrees of effect. These degrees differ from one person to another because we are not the same in everything. We have different talents and we grew in different environments with different cultures and different life experiences. All these factors are responsible for the formation of our life values which affect our response degree to each hook. So, everyone will be hooked to a field that is suitable for his talents and his capabilities and that is why you will be succeed in what you are interested in. So, don't choose other one's choice because what hooked other may not be suitable for you and vice reverse.

These different hooks, that are responsible for choosing of a specific scientific area, may be:

- **A book that we have read**
- **A teacher from our previous study stages**
- **Someone that we have met before**
- **A life situation occurred to us or to another one**

So, to find your interest, you should remember your early years of life. What was your favourite book, field, subject, teacher, hero, etc.? All these things can help you to recognize your interest. If you still don't find it, you should begin to read widely in science until you find your hook.

2.3. Find Your Guide....

"An expert is a person who has made all the mistakes that can be made in a very narrow field."

Niels Bohr

After the determination of your interested scientific area, your next step is to begin to acquire more knowledge in this area. Acquiring knowledge in a new specific scientific area needs a guide who is already specialized and expert in this area. It is very important to find an expert and a specialized instructor. Entering a new area of science is like entering a new city that you didn't enter before. When you want to find a specific place in this new city, you will need someone guide and help you to find your place and without him you will spend a lot of time and effort and you may not find what you want at the end. Also in science, you cannot understand and be specialized in any new area of science without the help of a specialized and an expert instructor who will be your guide in your journey in this area and will help you to find what you want and will save your time and effort. It was said that: *"who enter science alone will leave it alone"* which mean that if you try to get knowledge in a new scientific area without an instructor you will leave it without acquiring any knowledge.

Formal Education

Nowadays, Education process is completely different from the past. In the past, when someone wants to get specialized in any area of science, he began to search for distinguished scientists in this field who can help him to learn what he wants. But now it is completely different. There are formal education systems and these education system is divided into different stages begins by primary stages and intermediate stages and high school then the beginning of undergraduate then postgraduate studies. So, the process of searching a guide is completely different. Now, before graduation stage student are involved in definite programs with specific teachers or lecturers. These academic programs, either undergraduate or postgraduate, represent the role of the guide in this field at this stage. So, here you should choose to enroll in a program in your interested area.

Supervision

The ability to freely choose the scientific guide is begun after the graduation stage. In the postgraduate stage, during doing the PhD that is the stage of direct choosing of your guide that will be the PhD supervisor. Finding a good PhD supervisor is very important because he will guide you in your studies and you will work with him for years. You should search for a good expert PhD supervisor in your interested area. You can find him by searching about his previous work and reading his previous publications. You should try to find a PhD supervisor who not only a good expert in your interested area, but also good at interpersonal relationship level. This is an important point because you will work with him for years and there must be good relations between you and him. You can know information about his interpersonal relationships from the previous students.

Also, there is another guide type which is called a mentor. This guide is not your PhD supervisor, but he is someone who can give you some advices in your research way. You don't need mentor who is an expert in your interested area, but he should have a good experience about the research environment and the science way. He will support you with advices on your research way or when you face any troubles.

Ask for Help

The last point about finding a guide is to ask for help. This is very essential to succeed in your science way. You may face many difficulties in your research way and you should be able to ask for help. Some people avoid asking for help because they may be afraid of rejection or any other thing. Don't be afraid, you have to know that successful people will be happy to help you if you ask them. It's a Human trait, we are very happy when we deliver our experiences and advise others. Also, successful people have acquired great experiences during their life and they will be very happy to give you advices and to help you by delivering their experiences to you only if you ask.

In The End

You should try to find a good PhD supervisor in your interested area and a mentor who has good experiences about the scientific research way. And the most important thing is to not be afraid to ask questions.

2.4. Self-Education Skill….

"Teachers open the door, but you enter by yourself."

Anonymous

Your formal education or your guide will only clarify the map of your scientific area and supply you with a torch to enlighten the way, but you have to move to your specific destination by yourself and your way to this is self-education. Formal education is not enough to you to be a good scientist. Formal education can't supply you with all you need. Also finding a guide to help you in your chosen scientific area is very important but it's not enough. You need to do some work by yourself.

Importance of Self-Education

You must have the skills of how to get any information that you want. This skill will help you to increase your level of knowledge about your chosen field. You need more knowledge than those you have got from your formal education to be an expert in your field and to do your researches in an excellent way. Also, there are always continuous changes and progression in knowledge and it's very important, to be up to data, to not only depend on the knowledge that you get during your formal study or that your instructor gave you. Your formal education and your guide will supply you with the basics that can help you to know or to enter any site in your field. These basics represent the keys that you will use to move easily to get the needed information and to understand it. These basics will help you to understand the scientific terms in this field and to get a general view of this point and its relations to other points you know in your field.

Self-education is important in three issues.

- **First, it will help you to be able to find any information you may need during your research work.**
- **Second, it will help you to still up to date with the new findings in your field.**

Steps of Self Education

Self-education skill is divided into two main steps.

- **First, determine exactly what you want**
- **Second, choose the sources**
- **Third, find the sources**

Determine exactly what you want

First step, you should determine exactly what you want. You should determine your goals and what the level of details you want to know. The details' level of knowledge is different according to why you want it. You may want to know general view, more details or to use it in a practical way.

Choose the sources

The next step in self-education skill, is to choose the sources. This step should be taken very carefully. There are different sources that you can use in self-education such as;

- **Book chapter about what you want**
- **Books about what you need**
- **Review articles**
- **Online video lectures**
- **Online courses**

If you determine exactly what you want, it will be very easy to select the appropriate source. Based on your level of knowledge you should determine where to begin. For example, if you don't know anything about the topic, you should first take a general view. This general view could be taken from a book chapter specifically for your interest topic. Because when there is a book chapter only on a topic, it will give the main headlines about this topic. These headlines can help in forming a general view about the different dimensions of this point. If the whole book is on your interesting topic, it will give you more details. This book will more suitable for who has some knowledge about the topic of interest.

Another important source is the review articles that can provide a general review about your topic? These articles are very helpful because their information is supported with its citation which you can check to get more details or other references about the point you need to know.

Another current important source is video lectures on YouTube or other internet websites. Official websites for universities and scientific organizations provides good lectures about specific topic and some provide complete courses.

In self-education process, you should put the plan of study by yourself. You should move from what's general to more detailed points. What you want, where to begin, when to work and how to evaluate, these are really important points that you must consider.

Find the sources

There are many ways to find the sources of what you want to know.

- ***First Way***

You can check the reference list of a related book or article. Once you have found a book on your topic, it's really a great chance to find other sources through checking the reference list.

- ***Second Way***

You should ask the experts about what you want to help you how to find the appropriate source. Asking for help is not only important in getting sources, but also when you find something that you cannot understand.

- *Third Way*

The World Wide Web is a very important source today to find the sources that you want. You can use the scientific websites that contain the current researches in your field or about the point that you want. You can easily search the databases of these websites for what you exactly want. Most scientific journals have websites with databases about their published articles that you can search it. Also, there are different websites that provide online courses. You may find course about what you want in these websites.

In The End

Self-Education skill is very important for you in your way of science. You should have this skill. You should know how to get any information you want, to increase your knowledge and be up to date with the recent researches in your area of study.

2.5. Find A Position....

"Research is four things: brains with which to think, eyes with which to see, machines with which to measure and, fourth, money."

Albert Szent-Gyorgyi

Nowadays, scientific research is different from the previous periods. It is very expensive work; it depends on different advanced instruments and equipment. To do any experiment you need facilities and experiment instruments which hardly to get by yourself. You can get these scientific facilities through being affiliated with a scientific organization and by getting a position in it. So, it's very important to get a scientific position to do your scientific work.

Position in a scientific research organization not only helps you to get the instruments and equipment that you want but there are other important advantages. It will provide you with a scientific environment that will increase your focusing and concentration in your research work. This environment contains scientific discussions with other researchers, attending of scientific lectures, be up to date with the recent discoveries and the current big questions in your scientific field. This scientific environment will help you to be very enthusiastic in your work. Also, scientific position provides a financial support that will help you to spend your time freely in science. The life of the scientists and scientific researchers is similar to the normal person's life. They need money for their life and they need to spend part of their time in working to get this money. If they have a scientific position they will get a salary for their scientific work. So, if you get a scientific position you will not need to spend some of your time in working for money. You can save that time to spend it in your scientific work.

Scientific Position Criteria

To get a position in an excellent scientific research organization, you need to express that you are qualified for this position. To be qualified for a scientific position is to have three things; **interest, experience and abilities to do the tasks of this position**. You can express your qualifications through different methods.

The ways to express your qualifications

- **Recommendations**
- **Publications**
- **Certifications (PhD – Master)**
- **Work experience**
- **Books**
- **Scientific activities (Lectures, Conferences, Workshops, ..)**

Difficulties in Getting A Scientific Position

Although position is very important, some scientists find some difficulties to get one. The eligibility criteria were not compatible with them. But, they didn't surrender and tried to find another creative way to express their qualifications. One of the interesting examples in finding a scientific position is Michael Faraday. Although Faraday didn't complete his formal education, he was able to express his interest, experience and ability to work in the scientific field. Faraday was very interested in science. He was searching for a position that can help him to follow his passion for science and to work in what he loved. Faraday was working in the bookbinding trade. During his work, one of the costumers was satisfied with his work. This costumer gave Faraday a reward for his work. This reward was a ticket for a lecture by Sir Humphrey Davy at royal society. Faraday was very thankful for this reward. This ticket will be the first step toward his dream. On the day of the lecture, Faraday went to the lecture hall very early to get a closed position to hear and clearly see Sir Humphrey Davey. Faraday began to attend all the lectures of Sir Humphrey Davy. He noted down these lectures in his notebook. Then he studied these notes very carefully and put it in a thick book. Faraday presented this book to Sir Humphrey Davy and asked him to read it in his leisure time. After a period Faraday met Sir Humphrey Davy and asked him for his opinion. Sir Humphrey was very impressed by the book. Davy. Faraday asked Sir Humphrey for a job in his lab. After several months Faraday got the job and began his journey in science. So, if you find any difficulties to meet the scientific position criteria, you should find your own creative way to express your qualifications.

Scientists without A Scientific Position

Some people may search for getting a scientific position even if it's not related to their interest. They are not in the right way. Position is important for helping you on your way to great discoveries, but it's not a goal for itself. First, you should choose your interested area of science, then begin to get the qualifications needed for working on it then get a position. If you cannot get a position, you don't have to end your relation with science. There were other scientists that didn't have a formal scientific position and made remarkable discoveries that changed our view to our world such as Antonie van leeuwenhoek and George Mendel. Also, Einstein began his great scientific discoveries when he was working in the Swiss Patent Office in Bern. He tried in the beginning to get a scientific position, but he failed. In June 1902, he received the letter that he had been waiting for. This letter was a positive answer to his application to be a technical assistant level III at the federal patent office in Bern. He worked there in the period from 1902 to 1909. He said about his job "It was a kind of salvation". He divided his time in this period into 3 parts, eight hours of work, eight hours of miscellaneous and scientific work and eight hours of sleep which he often used instead for writing his manuscripts. During this period, he didn't cut his relation with science and he published very important scientific articles that changed our understanding about our universe. Einstein began to attract respect with his published papers and in 1909 he was appointed associate professor at the University of Zurich.

In The End

Getting a formal position in a scientific organization is very important. You should try to have the qualifications that can express your interest, your knowledge, your experience and your ability to work, to get one. But remember, it's not a goal for itself. If you cannot get a scientific position, you don't have to end your relation with the scientific field.

3

Step Three:
Ask To Know

The first step for you to acquire a new knowledge is to ask a question. The great scientists have the skill of finding and asking questions and you also should have this skill.

3.1. The Questioning Process....

"Questions are the creative acts of intelligence"

Frank Kingdom

The researchers and scientists, all the time, not only want to know the available knowledge, but they also want to discover new knowledge by themselves. This journey of discovering new knowledge begins with a very important step which is asking a question. Asking a question, is your first step to know more. Our universe is filled with an infinitive number of secrets and the main goal of the researchers and the scientists is to solve these secrets. But to solve these secrets and to discover more about them, they firstly have to see them. **Asking a question is the end product of the recognizing process of the secrets of the universe.**

Asking a question is a very important skill. It is the way by which we admit our ignorance and express it. The ignorance detecting is an important factor that stimulates people to know more. When we want to tell a person something we always begin by saying (Do you know.. ...?) Then, if we find him don't know anything about this topic we begin to say what we want to tell him. But why we begin by saying "do you know?" , although we may be sure that he doesn't know anything about this topic. But, the importance of this question is to increase the recognizing level of the ignorance that he has about this topic and this stimulate his interest to know more. So, the most important advantage of asking a question is to recognize the degree of ignorance that we have. Because **the first step to know is to know that you don't know.**

So, it is very important to have the skill of asking questions to discover new knowledge. Because once you get a question you will begin to search for the answer immediately. This skill is divided mainly into two parts. First part, you have to recognize your ignorance. Second, you have to express this ignorance in a right way. You can recognize the ignorance points by improving your noticing ability and by the skill of finding the gaps of knowledge. And to express the ignorance in a right way, you need to have the skill of how to ask the right question. These two steps are the fundamental steps of asking questions and you need to have them if you want to go through the discovering journey because **without the ability of asking questions you cannot discover new knowledge by yourself.**

3.2. Observing Powers....

"Knowledge is the intellectual manipulation of carefully verified observations."

Sigmund Freud

Observing ability is a very powerful tool that is used by the great scientists to find interesting questions. Our universe is working with definite laws that control all the behaviors of its components. Watching these different behaviors may lead to the observing of a remarkable and an interesting phenomenon. Watching the nature and the environment around us and thinking about their different processes and activities, that happen to them or by them, stimulates our minds to ask many questions like why, how, what if, when and where. Answering these questions could help to understand the secrets of our universe.

This observing ability is an important intellectual character of the great scientists. Scientists can see what others cannot see; they have the ability to see in a different way what we think it's a normal thing and should happen in this way. They look differently at any normal phenomenon and try to know why it happened in this way, they have the ability to observe and detect any change in it.

Types of Observations

The Observation is not only to see something, but to detect anything by your five sense organs. To observe is to see with the eye, to smell with the nose, to touch by the skin, to hear with the ears or to taste with the tongue. There are many wonderful and amazing discoveries that were begun by a careful observation of the nature. These different observations could be categorized and classified into 4 main types:

- **Observing of something new**
- **Observing of an unexpected behavior**
- **Observing of a remarkable relation**
- **Observing of a definite pattern**

Observing of something new

The first type of the observations is to observe something new. Many of the great discoveries that opened new gates for better human life were begun by a simple observation of new thing that wasn't seen before. One of the most interesting examples is the discovery of the microorganisms by Antonie van Leeuwenhoek. Leeuwenhoek was the first one who saw the single celled organisms by his new microscope. Leeuwenhoek was a draper from Netherlands. He was using the magnifying lenses to examine the quality of the clothes. He was interested in making lenses of high magnifying powers that can enable him to see more clearly. He learned how to grind lenses and made a simple microscope. This microscope had a magnified power over 200 X. One day, he used this microscope to examine drops from lake water. He began to see remarkable things within these drops. He couldn't draw these remarkable things by himself and hired an illustrator to draw it. He accompanied these illustrations with written descriptions in a letter to the royal society in 1674. He wrote *"Passing just lately over this lake, . . . and examining this water next day, I found floating therein divers earthy particles, and some green streaks, spirally wound serpent-wise, and orderly arranged, after the manner of the copper or tin worms, which distillers use to cool their liquors as they distil over. The whole circumference of each of these streaks was about the thickness of a hair of one's head . . . all consisted of very small green globules joined together: and there were very many small green globules as well."* He continued to examine and discover different new things by his new microscope, like blood cells and microscopic nematodes, sperm and other remarkable things. This is the first type of observations, to observe something new.

Observing of an unexpected behavior

In September 3, 1928 a Scottish microbiologist Alexander Fleming was in his lab in the basement of St. Mary's Hospital in London, returning from holiday. He began to sort through petri dishes containing colonies of *Staphylococcus*, bacteria that cause boils, sore throats and abscesses. He found that a Petri dish containing *Staphylococcus* plate culture, he mistakenly left open, was contaminated by blue-green mold. He examined this dish and observed unexpected behavior. He found something remarkable that the area around the mold was clear without bacterial growth in it. Why *Staphylococcus* colonies didn't grow in this area. He began to think that there must be a compound secreted by this fungus that prevented the growth of the bacteria. He then purified it and grew a pure culture and discovered it was a *Penicillium* mold, now it's known as *Penicillium notatum*. Fleming found that his "mold juice" can kill a wide range of harmful bacteria, such as streptococcus, meningococcus and the diphtheria bacillus. This was the discovering of the antibiotics, which is one of the most important discoveries that saved the life of many people. This great discovery only began with a simple observation of unexpected behavior.

Observing of a remarkable relation

In 1889, Oscar Minkowski and Josef von Mering were working in Strasbourg University. They were trying to investigate the role of the pancreas on the digestion process. They removed the pancreas from a healthy dog. A few days later, they noticed that flies were swarming around the dog's urine. When they analyzed the urine they found sugar in it. They realized the relation between the pancreas removing and the diabetes. This observation led after that to the discovery of the insulin, which was isolated by researchers at Toronto University. These great discoveries of the insulin and the role of the pancreas in diabetes were only begun by a simple observation of a remarkable relation.

Observing of a definite pattern

In 1582 Galileo Galilei discovered an important property of the pendulum that helped in using it as timekeepers. This property is called "isochronism". This discovery began with an observation of the motion of the swinging chandelier in the Pisa cathedral. He noticed that the swinging chandelier has a constant period to swing back and forth even when moving at different angles. When he went back to his home, he repeated these observations and performed different experiments. He discovered that the period of the pendulum is independent of the amplitude, width of the swing, the mass of the bob, and proportional to the square root of the length of the pendulum. After that, these observations lead to the invention of the pendulum clock and other different applications. This type of observation is to observe a definite pattern. Definite pattern is something that is always repeated in a specific manner or has definite shape or any remarkable pattern.

How to See What Others Didn't See

To be a good scientist, you should be a good observer. Thinking about what found in nature and why it happens in that way is an important intellectual character of the great scientists. Although the observing processes occur accidentally, the ability to observe could be increased. There are different ways that could help you to increase your ability to observe remarkable things.

- **First, begin to look in a different way.**
- **Second, prepare your mind.**
- **Third, find new scientific tools.**

Begin to look in a different way

To be a good observer, you need to have the ability to see the ordinary things around you in a different way and think about them differently. Albert Szent-Gyorgyi said that *"Discovery consist of seeing what everyone has seen and thinking what nobody has thought"*. Watching what happens around us is the key of the door to the great discoveries. All the activities and processes around us contain many messages about the secrets of life, the universe and the laws that control them. They only want a keen person that can see them and try to understand their meaning. Look in a different way and think about what happen around you and begin to ask why and how and what if and when and where. **Use this to ask interesting questions in your area of science.**

Prepare your mind

Observing ability is the ability to receive messages from what surrounding us about how our universe works. To be a good observer is to be able to see and to understand these messages. To get this ability you should have a prepared mind. Louis Pasteur said *"Where observation is concerned, chance favors only the prepared mind"*. Preparing your mind will help you to distinguish normal from abnormal and this is the most important advantage of it. This process of recognizing normal from abnormal could be only done by a specialized person who has knowledge about this topic. The same events could happen in front of many people, but not all of them can see it well. No one can differentiate between what's normal and what's not normal in a specific area if he doesn't have knowledge in this area. So, that's why preparing your mind will increase your chance to observe and your attention in different ways to many things that occur around you.

Find a new scientific tool

During watching of the nature you may find something new. Our sense organs have definite limits for detecting what occurs around us. There are many things and many events are around us that we cannot detect by our sense organs because they may be above or under our sensing ability range. The strength of our sensing ability could be increased by using assistant instruments that increase our sensing range. Finding a new scientific tool can help to notice remarkable phenomena and observe what nobody have ever seen before.

3.3. Find The Knowledge Gaps....

"Readers are plentiful; thinkers are rare."

Harriet Martineau

Finding the knowledge gaps is another excellent method for asking interesting questions. Our knowledge about any topic is incomplete and in a continuous progression all the time. There is always missing information that is needed to be known. This missing information represents the gaps of knowledge. These gaps lead to the arising of questions that needed to be answered. Every branch of human knowledge was resulted from the accumulation of the information that was discovered by the great efforts of the distinguished researchers and scientists. They are adding, and completing the work of each other.

When we become interested in a specific field, we begin to increase our knowledge in this field by learning more about it. During this learning process, we may find missing information that needed to be answered. By finding the answers of these gaps we begin to add to knowledge and this is how knowledge is progressed by every scientist. Every scientist detects gaps of knowledge and tries to find them.

How to Find The Gaps of Knowledge

The skill of finding knowledge gaps is very helpful in asking questions and in knowledge progression. But, how can we find the gaps of knowledge? The ability to find these knowledge gaps is varied from one to another. This ability depends mainly on the knowledge receiving manner which is varied between us. Some people can easily believe anything that they read or listen or watch, but on the other hand, there are other people who wanted to examine and to be sure about any information they learn and cannot easily accept anything. We can classify the different manners of knowledge receiving into main four types;

- **Learn to understand**
- **Learn by complete story**
- **Learn from different sources**
- **Learn by verification**

Learn to understand

The first knowledge receiving manner is to learn to understand. Albert Einstein said: *"Any fool can know. The point is to understand".* In this method you should try to know not only simple information about something or a specific topic, but you should try to see all the dimensions and details related to this information and the different interpretations of the meanings of this information. For example, if there is an information that A has an effect on B. Don't simply accept this information without asking why "A" has an effect on "B" and how this effect occurs. You should try to see the different events and processes responsible for the causing of this phenomenon and try to find the whole information and the whole picture of it. This knowledge receiving manner leads to the finding of different gaps and missing roles that needed to be understood.

Learn by complete story

The second type is to learn by the complete story. Lord Acton said: *"There is nothing more necessary to the man of science than its history, and the logic of discovery...: the way error is detected, the use of hypothesis, of imagination, the mode of testing".* Every scientific discovery has a story behind it. This story includes what was the problem? How it began? What're the sequential events that lead to the discovering of the solution or this knowledge? It is a way to follow the information from its first moment when it was a question with no answer. How this question arose in the mind of the scientist and what are the different activities that lead to finding of the answer. Learning new knowledge through knowing these stories, is a very powerful tool to reach the thinking levels that helped in finding this discovery and this raising in the thinking level could allow you to add to this knowledge through finding missing parts that was needed. In this type of learning you may ask questions from type "What if" to expect what might happen if the studies went in this way rather than that way. In this type of learning, you can see the information more clearly. For example, Instead of knowing that *"A"* has an effect on *"B"* you have to ask how they found that. By knowing the story you may find the missing parts that needed to be known. This manner of receiving knowledge not only helps in a clear understanding of this information, but it also helps in getting good experiences about how similar problems could be solved. And this experience could help in understanding and solving many scientific problems that may face you in the future. So, learning by complete story has many advantages.

Learn from different sources

The third type is to learn from different sources. During your learning and acquiring of the new knowledge you may find different sources for the same topic. You should begin with one of them, then examine the other sources for any additional information or other explanations that may help you to understand the topic more clearly. Examining different sources for the same topic could help in understanding the topic from different points of view. This may help in finding other missing information that needed to be known or you may find that there are no enough explanations for a specific point and this could lead to the arising of different questions.

Learn by verification

The last manner of receiving knowledge is to learn by verification. In this method you shouldn't believe everything you read. Many of the facts that were thought as completely true were found to be not true. Alhazen said: *"The duty of the man who investigates the writings of scientists, if learning the truth is his goal, is to make himself an enemy of all that he reads, and attack it from every side. He should also suspect himself as he performs his critical examination of it, so that he may avoid falling into either prejudice or leniency".* What you have learned, is the conclusions that were obtained through the works of the previous researchers and scientists. These are trials from them to explain what are happening in the universe. These trials were done in conditions that may lack some of the advancement that are presented today or in lacking of knowledge that are known today. The reexamination of these findings may lead to the discovering of new findings. This reexamination could be done by relating these findings to their relatives' points that are based on it and asking what if these findings were true and what if these findings were wrong and what are the results of these two conditions? Also, this verification could be done experimentally by repeating the work of others and examine what will happen. Celia Green said: *"The way to do research is to attack the facts at the point of greatest astonishment."*

So, the process of learning by verification begins when you receive a new information, either it was by reading or by listening or any other method. First step is to completely and clearly understand this information and what does it mean? Then start to imagine what the different probabilities that may occur if this information were true. Then begin to imagine the different opposite probabilities that could be resulted if this information was not true. And during thinking in this way, many questions may arise and these questions like what if.

In The End

To ask interesting questions you should begin to choose a suitable knowledge receiving manner that can help you to detect the gaps of knowledge.

3.4. The Questioning Minds....

"The scientist is not a person who gives the right answers; he's one who asks the right questions".

Claude Lévi-Strauss

Asking questions is a very distinguishing character of the great scientists. They were characterized by a questioning mind that make them asking all the time and about anything. Their ability to detect their ignorance was very high. And they also had the ability to express the question in the right way. As I mentioned before that the questioning skill is divided into two steps. First, you have to recognize your ignorance, then you have to express this ignorance in a right way. It is important to express your ignorance in an appropriate way that can help you in finding the answer or making the process of finding the answer very easy. Scientists have this skill of asking questions in a right way.

The Core Question

Our ignorance can stimulate us to ask many questions. But not all the questions are on the same level or the same degree of importance. The Importance of the question is based on its ability to solve a large part of the problem. You can ask many questions about the problem, but there is a question that its answer is enough to make the problem clearer and could be solved. This is the core question or the key question that could help in solving most of the problem. The core question can attack the heart of the problem and its answer can destroy the rest of our ignorance about this problem.

Find The Core Question

The core question will easily direct you to find the right answer. But, how can you find the right question. To find the core question of the problem you have to divide and classify the ignorance into layers. Each layer is originated from another layer until you find the central layer of ignorance that from which the rest of the problem was originated. There is always a central layer of ignorance in any problem. It is existed because the presence of the relations and dependence between the different parts of the problem. You should try to determine these basic relations that links between the different parts of the problem.

In searching for the right question you search for what is the point that if it was known, it could help in solving most of the problem? Rather than asking many questions about the problem and trying to solve each one alone, you only need to find the relation between these different questions and ask one question about the heart of the problem. So, you should try to find how all parts of the problem are linked together or what the main idea that can affect these parts.

The ability to ask a right question is to search for the links and to relate different problems until you find the central layer of ignorance. The Link is something in common or can lead to the formation of this problem. This link may be the place or site of work, time, mode of action, etc.. . There are two ways for finding the central layer or the core question by searching for the link.

- *Divide*

In the first way, divide the existed ignorance into parts and treat each part as a separate problem and try to find how each one of these problems is originated and what the links between them are.

- *Enlarge*

In the second way, enlarge the existed ignorance by relating it to other problems of the same type or related somehow by a link.

In The End

The ability to find the right question is very important to easily solve the scientific problems that you may find. **So, you not only have to ask questions, you need to ask them in a right way.**

4

Step Four:
Find A Solution

The searching of the answer of a question is the main goal of the scientific research. It is a very special moment when you solve a scientific problem.

4.1. Discover New Land….

"The process of scientific discovery is, in effect, a continual flight from wonder."

Albert Einstein

Now, you may have a scientific problem and you want to begin your research journey to find the solution. For me, I think that finding new scientific discovery is like finding new land for the mankind. Your journey to find a solution of a scientific problem is like the journey of the sailor in his sailing journey to discover a new land. The similarity is not only in that both of you and the sailor work to provide great benefits to mankind, but there are also similarities in the tools that are used.

Prepare Your Tools

As the shipmaster needs some tools to succeed in his mission, also you do. First, he needs a ship to carry him in the sea and need an experience about how to use it. Also, he needs a map to use it in the determination of his site and his destination point. Also, he needs a compass to determine his right direction and an experience about how to understand the signs of the sea. You need similar tools in your researching journey. You also need a ship, an experience about what you do, a map, a compass and an ability to understand different signs.

- **Your sea is your chosen field of study.**
- **Your destination is the shore of the truth and to find the answer of the question.**

- Your ship that will carry you is your basic knowledge that you have about this field.

- The map that you will use to know the way is the previous studies of the other scientists in the same field.

- The compass that will help you to know how to move to your destination is your thoughts and expected solutions for the problem that are based on your intellectual analysis of the available information about the problem and your previous experiments.

So, you should choose your sea, prepare your ship, find your map and your compass and pray to God to help you reach the shore of knowledge and discover new land for all the mankind.

4.2. Genius Minds....

"I don't divide the world into weak and strong or successes and failures. I divide the world into learners and non-learners."

Benjamin Barber

Some people think that scientists are a special group of mankind with special mental abilities that are not existed at anybody else. These thoughts are not true. Scientists don't have special mental abilities. Their minds are like yours. The difference between them and you is that they use some mental capabilities that all people have but not used by all. Your brain has fabulous and amazing capabilities that can enable you to find amazing and wonderful discoveries.

Change Your Mind

In a study at Stanford University, Professor Carol Dweck found that people are divided into two groups; the fixed mindset and growth mindset. People with the fixed mindset, think that the intelligence is a fixed trait that was born with us. Instead, people with the growth mindset, think that intelligence is a quality that can be changed and developed. People with the fixed mindset, think that the intelligence is fixed and permanent. They think that who are born smart will still smart and who are not will still not smart and they cannot be changed. Instead, people with the growth mindset, think that there is always a way to grow in anything they want and they have the ability to acquire the skills that they want.

Fixed mindset is a psychological trap that prevents success. When people with the fixed mindset examine themselves in any situation and find that they are not good in it, they judge themselves that they are not smart like others and they will not achieve anything in their life. And If they succeed in it, they will think that they are special with high degree of intelligence and this will make them think that they don't need to practice more or to work hard. This also will make them fail. Dweck found that people with the fixed mindset always fear to challenge themselves and disengage from solving any problem.

Many of us may have these wrong thoughts. In childhood, these thoughts could be built when someone said to you that you are smart or you are very talented. Then you believed that you have special capabilities and these capabilities are enough for your success and you don't need to work hard. Or when someone said to you that you are not smart enough and you will not success in your life. Then by believing these thoughts, you thought that your thinking capabilities will not be changed and you will surely fail. These thoughts might be built inside you by your parents or teachers when they tell you that you have intelligence or not. This is a wrong message.

Instead, you must know that you have a highly manipulated brain that can grow if you practice more. Hard working and learning can help you to increase your mental abilities. You should change your mindset. You don't lack mental abilities, but you only need to learn more skills. Take care, you may see yourself as a non-brilliant mind it's not true. The main problem is that you don't know the amount of your mental abilities and you lack the knowledge to use it. If you know how to use your thinking capabilities and begin to use it, you may find yourself a different person. Do you think that Einstein was born with his high mathematical skills. When he was born, he didn't know even that $1+1=2$. But he developed his high mathematical skill by learning and through hard working.

Develop Your Ways Of Thinking

No one can ignore that the great scientists, who made the wonderful and amazing discoveries, have very good thinking skills. But, you should know that they acquired these skills and were not born with them. Scientists were not specialized in creative thinking skills. Nowadays, there are many courses about the skills of problem solving. But that wasn't available to the previous scientists. So, how they acquired these different mental skills. They developed their ways of thinking either through practicing or by taking it subconsciously from other scientists through reading other scientists' works. Reading in science will make you subconsciously acquire the problem solving skills from other scientists work. Continuity of thinking is also a powerful way to increase your mental abilities. When Newton was asked how he discovered the law of the universal gravity, he said that *"by thinking on it continually …. I keep the subject constantly before me and wait till the first drawings open slowly, by little and little into a full clear light"*. Continuity of thinking about any interesting problem increases the mental practicing which increases its thinking ability. So, you don't need to be professional in thinking, but you have to be passion about what you want to know and by your readings and hard working your mental abilities will be increased.

4.3. Solve The Problem....

"Science is a way of thinking much more than it is a body of knowledge."

Carl Sagan

Once you have find an interesting question, either it's yours or not, your next step is to find its answer. During searching for the answer of a specific scientific question, actually you are trying to understand more about a particular process or a natural phenomenon. Any particular process or natural phenomenon has two parts; an unknown part which you are trying to find and also have a known part that is the available information about this problem. During the process of finding the question's answer, your goal is to increase the known part about the problem through decreasing its unknown part. So your main goal is to understand more. To understand any natural phenomenon you should try to find what are the different factors responsible for it? What's the role of each factor? And what's the mechanism through which these factors do their roles? So, scientific problems are divided into different types according to the part that you try to know. You must first determine what the known parts and what the unknown parts and classify them and relate each of which to its type.

The process of solving any problem is begun by two main steps.

- **First, gather information about the problem.**
- **Second, begin to reorganize this information.**

These steps will help you to increase the clarification of the problem and will help you to understand more about the particular process of interest.

Gather Information about The Problem

Your first step on your way to understand the problem is to read more and to gather more information related to the problem of interest. You may have some knowledge about the problem, but you should also find more information about it. Science is a collaboration process that based on the efforts of different scientists and you should read and find the available information from the other scientists' works and studies related to your problem. Isaac Newton Said: *"If I have seen further it is by standing on the shoulders of Giants."* Nowadays, the process of finding the available information about your interesting problem is called the review of literature. I talked before about the self-education skill and how it is very helpful skill to find any information you need. You should have this skill to gather information about the problem.

When you search to get information that you want, begin by asking questions and expecting their answers. Then, you have two probabilities. You may find it either as you expected or not. These expectations help in increasing your awareness with the details of the problem and also may help in finding new ways that weren't investigated before.

Reorganize The Available Information

The Finding of the answer of the problem mainly depends on the ability to reorganize the available information. Martin Fischer said: *"Facts are not science — as the dictionary is not literature."* In order to completely understand the problem you need to use the known information in finding the unknown parts. This transfer from known to find unknown begins by the process of information reorganization. By this step the problem will be clearer and you can determine accurately the missing points. There are different strategies that could help you in reorganizing the available information;

- **Think by analogy**
- **Think by pictures**
- **Search for the relations**

Think by analogy

It is a very helpful strategy to understand more about the problem. In this method you try through the available known information about the problem to explain its points by using a well understood example that seems similar to your problem in some points. There are many thoughts that consider that the whole universe was created on the same laws and rules. So, finding something similar to your problem may help in finding the unclear side of your problem by considering that they will act in the same behavior.

The most famous example for using analogies in science is the using of the solar system to explain the structure of atoms. You can use different analogies for the same phenomenon because it may resemble one phenomenon in part and other phenomenon in another part.

Steps of thinking by analogy:-

1. **Choose a well understood example that has some similarities with the studied phenomenon.**
2. **Aligning different characters and parts of this example with the studied phenomenon.**
3. **Try to make new inferences through comparing between the analogy and the phenomenon that could help in understanding the problem.**

Think by pictures

It is a famous way of thinking that was used by many distinguished scientists like Faraday and Einstein. In this method, you can transfer all the available information and all the imaginations of your mind into pictures on a blank paper. This will help you to strongly refine your thoughts. And the using of the visual materials can stimulate other parts of your minds which could help in understanding more about the problem. Reorganize the information in more than one picture or way, then put the pictures (side by side) together and determine the missing information and try to expect it.

Search for the relations

Reorganize the different problem's information based on the different relation between it. Search for the different relations between the available information. These different relations may be similarities or differences based on the characteristics related to the studied phenomenon such as time of action, site of action, mode of action, its function, its properties, its effect and so on. Reorganize the available information according to these different relations.

Consider These Notes

There are notes that must be considered during your understanding journey

Treat the problem like a puzzle game

Think about your problem as a puzzle game and you are trying to organize its different parts, which are the available information to form the whole right image. But here, there are missing parts so you should reorganize the parts again and again to determine the missing parts.

Try to find the different possibilities

If the problem has two different opposite possibilities and its answer may be one of them, begin to search for a third possibility that can mix or merge between the other two possibilities. This new possibility may also have another opposite possibility, find it, then find the merge between them. For example, if there are two possibilities that "**A**" may occur before "**B**" or "**B**" may occur before "**A**". You should think about the other possibility that "**A**"&"**B**" may occur at the same time. This way of thinking could increase the possible answers.

Compare between unrelated things

Comparing between your point with unrelated things and finding the similarities and differences can help in more understanding of the problem through finding other unclear sides that could help in finding the answer.

Be flexible

Remember, when you are transferring from known to unknown to find the solution you should be flexible because **it's an unplanned journey.** You are moving from point to another based on the progress of the researching process and the available information that will determine your direction and your next step.

Rephrase the question again and again

An important step to find the answer is to rephrase the question in different ways and from different sides. Rephrasing the question is a continuous process during all the steps of the researching process. Reorganizing the available information again and again may lead to the need of rephrasing the main question again to find the right and the more accurate one. Try to ask many questions about the problem by using different question words such as who, what, how, when and where. Also, you should begin with finding the answer to these questions if it in the available information about the problem. Another way to rephrase the question is to divide the problem into parts and ask small questions and try to find their answers.

Don't overwhelm yourself with information

An important note about dealing with the previous information of your problem point is to not overwhelm yourself with the information. The high amount of information may be somehow an obstacle to solve the problem. Daniel J. Boorstin said that *"The greatest obstacle to discovery is not ignorance? It is the illusion of knowledge."* Try only to find the required information not all the available information related to your problem.

Follow the ripping of the ideas

During the organization of the available information, you may find other expectations for the answer of the problem. You may also get new idea related to your problem. You must learn how to follow the ripping of your ideas. Ask questions from types "what if?". What if this idea is true and reorganize again and again and examine the logical validity of your idea.

4.4. The Eureka Moment....

""When I am working on a problem I never think about beauty. I only think about how to solve the problem. But when I have finished, if the solution is not beautiful, I know it is wrong."

Buckminster Fuller

Do you have a problem and you haven't found the solution yet? – Did you do every possible thing that you can to get the answer of the question you have and there are no results? Calm down and don't worry, my advice to you now is to move your thinking away from the problem and to think about any other thing. You can now take a break, go to sleep, take a shower, go to walk or do anything rather than thinking about the problem. But take care, the answer may pass in front of you in any of these moments.

In many cases ideas don't appear when you are searching for them. Did you lose something before and you tried hardly to find it without any result. Then after you stopped searching and began to calm down and relax you remembered its place or found it accidently. This is actually what happened to many of the great scientists during their journey of discovering. Many of them tried hardly to solve a scientific problem and made all what they can do for finding a solution. Then, the answer was appeared when they stopped searching for the solution and went to do anything rather than thinking about the problem. This moment when you find the solution of the problem is called the "Eureka" moment.

Archimedes' "Eureka!" moment

The story of Eureka began before 2200 years ago. Archimedes was asked by King Hieron II of Syracuse in Silcily to examine a golden crown that was made for the king. King Hieron II was suspected that the jeweller had substituted some of the gold with silver. So, Archimedes task was to determine whether some metal rather than gold had been used in making the crown or not. At this time, it was known that gold is denser than silver. It means that, the volume of a golden material will be less than the volume of a silver material of the same weight. So, the golden crown should have a specific volume according to its weight. He can measure its weight easily. The problem was in how he can determine the volume of the crown that had an irregular shape. Archimedes spent a long time in thinking to solve this problem with no results.

While he was in the bath, he noticed that water was splashed out in the moment he got into the bathtub. He realized that the amount of splashing water is related to his body volume. He thought that he can do this with the golden crown. If he submerged the crown in water, it would displace an amount of water equal to its volume. So, he could determine the volume of the crown that has an irregular shape. At this moment he leapt out of the bathtub and ran through the streets of Syracuse naked shouting 'Eureka!' 'Eureka!'. The word "Eureka" means "I found it". This was the "Eureka!" moment. From this moment, the word "Eureka!" was used to celebrate solving a scientific problem or finding any wonderful idea.

How to Get A "Eureka!" Moment

All scientists wish to find such a "Eureka!" moment. This is the moment at which they reach the solution and completely solve an interesting scientific problem. Archimedes story was replicated with many scientists. It began with finding a problem. Then they tried to do all what they can to find the answer and there were no results. At the end, when they stop thinking about the problem and turn their thinking away from the problem, the answer appeared. This is what happened to many of the great scientists. To have such a "Eureka!" moment you should know its characteristics.

First, do all what you can

The "Eureka" moment comes when the scientists turn their thinking away from the problem after a long period of thinking about it. No one can know when ideas will come. There is no definite time or place to find them. All what you can do, is to increase your receiving ability to get an idea. This receiving ability increases by thinking about the problem and trying to solve it.

Second, take a break

"Eureka!" moment always occurs when the scientists are in a relaxed state and turning their thinking away from the problem. This may be due to the effect of dopamine. I talked before that the increase in dopamine increases our creativity and it's produced mainly when we are relaxed. So, this could explain why ideas appear when we are relaxed.

Third, look around

In many cases "Eureka!" moment begins by noticing something which acts as the spark. This spark is an observation that inspired them to see the solution. This may be due to the laws that control the universe is widespread. You may face or see something that's worked or controlled by the same law that control your interested phenomenon. So, seeing it in a different form may explain the other form.

In The End

The "Eureka!" moment is an astonishing moment that all scientists wish to have. You should first do all what you can, to solve the problem. Then, if you didn't find the solution don't worry. Now, you can take a break and begin to think about another thing away from your problem. At this moment the answer may pass in front of you.

5

Step Five:
Verify Your Answer

Once you have proposed a solution, try to find a way to prove it. The way of verifying the answer is not less creative than finding the answer.

5.1. The Right Answer....

"In questions of science, the authority of a thousand is not worth the humble reasoning of a single individual."

Galileo Galilei

You should verify your thoughts and opinions to reach the truth. This is the next step after you have proposed what you think it's the true answer of the problem of interest. You should find a way to verify your suggested answer. The scientists and researchers are truth seekers. They search for it all the time. Their main goal is to find the true answer of a scientific question. One of the very important questions about this searching process is how they can ensure that their findings are the real and true answer or not. Scientists try to find the answer for this important question to be able to differentiate between what is science and pseudoscience. This need made them search for the rules that control the research process and finding of the true knowledge. These rules that control the process of scientific research now are called the scientific method.

The Difference between Science and Pseudoscience

Humans can acquire knowledge form different sources. Helmstadter (1970) identified 5 common methods of acquiring knowledge. The first method is **the tenacity**, which the acquisition and persistence of an idea accepted for so long. The second method is **the intuition** that is the gaining of knowledge without reasoning, interpretation or inferring. It is mainly depending on the feelings. I feel true, although I can't really tell you why. The third method is **the authority** which is the acceptance of the knowledge because its source is highly respected. The fourth method is **the rationalism** in which knowledge is gained only by reasoning and logical thinking. The fifth is **the empiricism** in which knowledge is gained from observations based on the sense organs.

These are the main different methods by which humans get their knowledge. But how we can differentiate between what's true and what's not. Not all knowledge could be considered as science. Knowledge should be examined by different ways to verify its correctness or falseness. Knowledge is divided scientifically into two main types. The first type is **Science** or Scientifically accepted knowledge which are proved claims, verified information and supported by scientific evidence. Although there's no certainty in science, scientifically accepted means that it's not completely true, but it has supported evidences. The second type is **Pseudoscience** which is claims seen as science, but they haven't scientific evidence that can support them. So, we need evidence to accept any knowledge as science.

Scientific evidence is the kind of thing that serves to either support or counter a scientific theory or hypothesis. There are different types of the scientific statements based on the strength level of evidence. First **the observation,** it's the recorded data which are acquired by the sense organ. It needs to be interpreted and explained. The second statement is **the hypothesis** that is an educated guess based on the observations and the intellectual analysis of the available information about the problem of interest. It's not strongly supported and needed to be tested. The third scientific statement is **the theory** which is an acceptable explanation for a particular phenomenon or related phenomena based on a repeatedly testing of a hypothesis or a group of related hypothesis. It's strongly supported scientific statement. The fourth scientific statement is **the law** which is a concise description of the relation between definite factors. It's always expressed by mathematical equations. It only describes the relation without explaining it. It's also strongly supported scientific statement.

Philosophers and Scientists

It's clear now that the scientifically accepted knowledge needs evidence. But how we can find the scientific evidence? It's a big philosophical problem. Scientific evidence is determined based on the established rules that control the scientific research process. It is necessary here to differentiate between the role of the scientists and the role of the philosophers in the scientific research process. The scientists are individuals who are trying to solve the secrets of our universe by asking questions and trying to find their answers. But, the philosophers are individuals who mainly concern about the rules that determine the validity of the research process findings. According to the philosophers, Knowledge's philosophy could be distinguished into ontology and epistemology. **Ontology** deals with the nature of reality. Epistemology deals with the relationship between researcher and research object. **Epistemology** mainly concern with what knowledge is and how it can be acquired. The rules that control how knowledge could be acquired is the scientific method.

Brief History of The Scientific Method

The scientific method has been developed for a long period of searching and developing and by effort of many great scientists and philosophers. This process of finding these rules was associated with the development of science itself. It is a continuous process and has begun since thousands of years. These are some of the important steps that had a great effect on the development of the scientific method.

Plato

Plato believed that humanity was born with innate knowledge of everything. And the learning process is unlocking of memories. He thought that pure knowledge could be gained by deduction alone without a need for observations to verify the rational knowledge. The deduction is a top down logic, in which the conclusion is drawn from general premises.

Aristotle

Aristotle didn't agree with Plato in his thoughts about the acquiring of the pure knowledge. Aristotle thought that empiricism is important in acquiring knowledge. Empiricism means to get knowledge by our sense organs. He proposed that the universal truth can be known from particular things by induction. The induction is a bottom up logic, in which a general conclusion is drawn from specific premises. But, although Aristotle considered the importance of the empiricism, his science isn't empirical in form. He didn't mainly rely on the experimentation or observation in getting pure knowledge. He believed that reasoning is the best way and used syllogism to infer new universal truths from those already established.

Islamic civilization

The golden age of the Islamic civilization was the beginning of the real combination between both reasoning and empirical methods. It was the first clear presence of the importance of the experiment in science. Muslim scientists used the experimentation to verify the correctness of the previous scientific theories. One of the best brilliant minds of this period of the Islamic Renaissance was Ibn Al-Hytham (Alhazen). He was a physicist who proposed definite rules which are a combination between observation, experimentation and rational argument, to find the pure knowledge. He is considered as the first free scientist because he developed a scientific method very similar to the current used one.

1. **State the problem based upon observation and experimentation**
2. **Formulating a hypothesis and testing it through experimentation**
3. **Interpret the data and come to a conclusion**
4. **Publish the findings**

Modern Science:

The scientific method continued to be developed and passed through different stages from the beginning of the European Renaissance until the current modern formulation of the scientific methods. These different developmental stages were performed by great efforts of many scientists. The modern formulation of the scientific method is called the hypothetico-deductive method. It's the current scientific way to find the true answer of the scientific problems. It's consisted of definite steps:

1. **Observation**
2. **Question**
3. **Formulating of Hypothesis:** it's a proposed answer for the question.
4. **Predictions:** based on the deduction of the formulated hypothesis.
5. **Testing of predictions:** by performing an experiment
6. **Collection of data**
7. **Analysis of data and drawing a logical conclusion**

After reaching the conclusion another prediction based it could be found and these different steps are repeated. The hypothesis is confirmed when true observational consequences can be deduced from it.

In The End

Finding the right answer is the goal of the great scientists. It's one of the most important things related to the scientific research, to know how we can reach the right and true answer. Many scientists all over the previous ages tried to answer this question and formulate the rules that could help in finding the right answer for the scientific questions. These rules were called the scientific method. Now, the modern formula of the scientific method is the hypothetico-deductive method which is a combination between both reasoning and empirical studies that help in verifying the suggested answers of a scientific problem.

5.2. Experimental Ways….

"An experiment is a question which science poses to Nature, and a measurement is the recording of Nature's answer. "

Max Planck

Experiment is the third step in your discovering journey. This is the next step after you have proposed an answer for the question. To reach the right solution of a problem, you have to verify your proposed answer. Experiment is your practical way to verify this suggested answer. To do any experiment you have to determine the hypothesis and the experimental way.

The Hypothesis

The first step in doing any experiment is to determine the hypothesis. The hypothesis is the rephrasing of the proposed answer of the problem in a way that enables the examining of its validation. In this hypothesis, based on the proposed answer and the available information, you predict the results of the experiment. There are some points that you must consider about stating a hypothesis. First, the hypothesis must be testable, which means there are observations, based on the correctness of this hypothesis, could be tested. Second, it must be falsifiable or be possible to be falsified, which mean if it is false there are a ways to prove that. For example;

Hypothesis A: "The red color is better than the yellow" color (This hypothesis can't be tested; so, it's not a testable hypothesis)

Hypothesis B: "There is a life on other planets" (This hypothesis can be tested; if we find any observation of a life on other plant we can prove its correctness. But it's not a falsifiable because there is no way to prove its falseness if it was not true)

Hypothesis C: "There is no life on any other planets" (This hypothesis is testable and falsifiable because it could be tested and there is a way to prove its correctness or its falseness. If we find any observation of a life on other plant the hypothesis will be false. So, the hypothesis is falsifiable)

So, the testable and falsifiable hypothesis must have observations that can assess its correctness or falseness. Then, if the testable and falsifiable hypothesis is tested and the results are significant, it can become accepted as a scientific truth.

The Major Ways of Experimentation

There are different types of experiments and these types are different according to the type of the problem. As it was mentioned previously, any scientific question is aroused in order to understand more about a specific process or any particular natural phenomenon. Any particular phenomenon is processed by a group of factors. Each factor has specific properties and a specific role in this phenomenon and these different factors' roles are performed according to specific sequence and organized according to the specific mechanism of action. So the problem mainly deals with one of the following questions;

- What are the responsible factors?
- What are the properties of these factors?
- What is the role of each factor?
- What are the mechanisms in which these roles are performed?

The main goal of the experiment is to produce a particular system, of determining conditions with controlled variables, to allow the examination of the validation of the suggested answer. According to the goal of the experiment, there are different ideas by which the experiment could be done.

The ways of factors determination

To determine the factors that control a particular phenomenon, you should first determine the suspected ones. This could be concluded from the available information. After you have determined the suspected factors, the next step is to ensure that these factors are actually involved in this phenomenon or not. This examination could be done in different ways

- *Mark the suspected factor*

In the First way, add a mark to the suspected factor through which you can investigate its effect either through determining its site of action or through determining its time of action or anything related to its action and from which you can find is it involved in the process or not.

- **Eliminate the suspected factor**

In the Second way, try to eliminate the suspected factor and compare the two cases of absence and presence of it to ensure is it involved in the process or not.

- **Increase the effect of the suspected factor**

In the third way, try to increase the effect of the suspected factor by increasing the presence of this factor and examining the result of this increase to ensure its effect on the particular phenomena.

The ways of studying the properties of the factors

Studying the properties of the factors that control a particular phenomenon is very important to determine how this factor performs its effect. Every factor has different properties and these properties could be classified into two main types. **General properties** are the factor's properties that do not have an effect on its role in this particular phenomenon. **Action properties** which are the properties related to the factor's functions in this problem. Action properties could be divided into 2 main types:

- **Built in properties such as; structure, weight, volume, length etc.**
- **Reflected properties which are the reflection resulted from the built in properties**

The investigation process is based mainly on the previous analysis of the available information about the problem. To determine the factor's properties you should have information about the properties that are related to the existence of this phenomenon or have an effect on this phenomenon. It is very important to narrow the area of research and try to investigate these properties.

In studying the properties of a factor you want mainly to know what this factor can do and how it can do that and what the receivers of its effect are. Studying the built in properties help in determining its reflected properties. Also, analysis of the reflected properties helps in determining of the built in properties. To study the reflection of this property, there must be a receiver that could be affected through this action of reflection and by its response degree the level of property could be determined. There are two main ways to study the properties of the factor:

- *Detect the response of a suitable receiver to the factor effect.*

To measure a specific property of a factor, you can study the effect of this property of interest (its degree) on a suitable receiver (affected) and detect the result of this effect (the response). As (temperature with thermometer) or (weight by balance)

- *Detect the response of the factor to the effect of another factor.*

Examine the response of the factor of interest to the effect of other known factor. Then, measure the results of the effect or the resistance of the factor of interest which mainly depend on its specific properties. For example, to study the rigidity properties of something you can affect it by other known factor to test the degree this rigidity.

Ways of studying the role of the factor in a particular phenomenon

For studying the role of a factor in particular phenomenon, you need to **study its real and actual effect in the process of interest** and to do that you need to know **(the factor, the effect, the receiver and the results)**

- **The receiver which is affected by the effect of the factor.**
- **The effect is a type of relation between the factor and the receiver**
- **The result is the response to the effect of the factor on the receiver.**

To completely understand the effect of the factor you can first suggest it based on the analysis of the available knowledge about its properties. Then, examine this effect on the affected receiver experimentally. You can examine it in its natural condition if it's possible to control the other factors and there is an ability to detect the results. If there is a problem in examining the effect in its natural conditions you can separate the factor and the receiver together and examine them alone.

The factor's effect could be received by more than one receiver. And it may affect these receivers in different ways. So, every receiver should be examined for its response for the factor effect. As the factor can affect more than one receiver also the factor could be the receiver of other factor's effect.

Ways of studying the mechanism of action

Finding the mechanism of action is an intellectual process more than it is an experimental one. It depends on the analysis of the available information about the factors that control the phenomenon of interest, their properties and their examined roles. Studying the mechanism of action is to study the sequence in which each factor performs its action. To completely determine the mechanism of action you should have detailed information about each factor. This information includes **(the factors' roles, their time of action, their site of**

action, their initiation conditions, their degree of effect, their
termination, their receiver and their results)

After the suggestion of the possible mechanism, it could be
examined experimentally by disruption of specific stages in its
natural case and examining the results of this disruption.

"What if" experiments

The importance of the intellectual analysis of the available
information about any studied problem is to narrow the
proposed solutions for this problem. If there is no enough
available information about the studied problem we can use
another type of the experimentation which is **"what if"**
experiments. **"What if"** experiments help in finding a raw
material (information) that will be used in the intellectual
analysis to understand more about the problem? That's like
throwing a stone in stable water, aiming to find something
valuable.

"What if" experiments are not performed to examine the
validation of a specific hypothesis, but it's an explorative
experiment more than it's a verified one. In this type, there is
no enough previous information to help in predicting what
could happen in a specific situation so this type of experiments
are done to gather more information that could help in
understanding the problem. "What if" experiment is done to
examine what will happen in the case of the existence of a
specific situation.

In The End

The research process of understanding any natural
phenomenon is continuous and needs the efforts of many
scientists to solve and understand its different problems. One
scientist may spend all his research time in studying only one
factor of this phenomenon and other scientist could help in
finding one of its important properties and another one could
determine its role and so on. This work could be done by
scientists from different disciplines and it may continue to
different generations.

5.3. The Experimentation Work....

It doesn't matter how beautiful your theory is, it doesn't matter how smart you are. If it doesn't agree with experiment, it's wrong."

Richard P. Feynman

After you have determined both the hypothesis and the way of the experiment, the next step is to do your experiment to get the results that will help you in examining your hypothesis. The experimentation process should be done very carefully and in an accurate way. To ensure this accuracy, there are some notes that you must consider during the experimental work.

Types of The Empirical Measurements

The empirical studies could be divided into two types:

- Scientific experiments: in which you try to mimic the conditions of the studied phenomenon with some changes that are performed by yourself
- Natural investigation: in which you try to investigate the natural phenomenon in its natural conditions and try to collect data in different conditions of it.

The Main Items of The Experiment

1. The problem or the question
2. The hypothesis
3. The way of the experimental work
4. Tools of experiment (Instruments and Materials)
5. Steps of action

The Main Principle of The Experimental Work

The main principle in any experiment is to examine the relation between an effector and a response based on this effector. To do this examination, you try to measure the effect of the variation in the level of the effector on a specific response and to control other not studied variables. So, after determining of the way of the experimental work, you should determine the following types of variables:

- **Independent variables: the factors that you want to study its effect**
- **Dependent variables: the variables that you want to examine its response**
- **Constant variables: the variables that are not studied in this condition and shouldn't be varied.**
- **Uncontrolled variables: the variables that cannot be controlled**

Experimental Tools

There are different types of the experimental tools based on the different stages of the experiment.

Establishing tools

These are the tools that are used in preparing and establishing of the experiment

Running Tools

These are the tools that are used in the proceeding of the experiment and through which, you can control the independent variable and could help in collecting the samples or online data (the data during experimentation work).

Detecting Tools

These are the tools used in measuring and detecting of the variations in the dependent variables. The main idea of these tools is to increase your sensing ability and to be able to detect the small degree of variation with high accuracy. Detecting tools could also be used in preparing the experiment and also could be used during the running stage to get data during the experimentation.

Criteria of The Experimental Work

Any experiment -regardless its type or its goal - must have some important criteria to be done in a right way:

Reproducible

Reproducibility is the ability of the experiment to be reproduced or replicated either by the researcher or by someone else working independently. And the same experimental results could be achieved if the experiment is replicated with different operators, test apparatus, or laboratory locations.

Unbiased

Unbiased means that there is no previous prefer that you motivate the experiment to prove it. This bias could appear in sample collection or in sample analysis.

Replication

The experiment should be repeated different times to eliminate the random errors

Control

When you examine the effect of a specific variable and get the results of it, to ensure that the results occurred due to this tested variable and not because any other one, you must do a control experiment.

In The End

The experiment should be performed very carefully and in an accurate way to get a good data that can be used in drawing a right conclusion.

5.4. The Right Conclusion....

It is a capital mistake to theorize before one has data."
Arthur Conan Doyle

The main goal of the experiment is to verify the hypothesis by testing the predictions that are deducted from it. Data are the observations and measurements that are collected during the experiment. You need to interpret these data to get a reasonable conclusion.

Steps of Data Analysis

The main goal of the data interpretation is to draw a right conclusion from the experimental data. There are different steps to draw a reasonable conclusion form the recorded data.

- First, recording of data.
- Second, examining the quality of the recorded data
- Third, finding the correct conclusion through logical thinking

First recording of the data

The first step in the process of data interpretation is to collect the experimental data in a right and accurate way. Scientists used a lab notebook to record the experimental data. The lab notebook is the researcher's notebook in which he records his hypothesis, thoughts, experiments, measurements, primary interpretations of the data and all the information related to his research work.

Lab notebook has many advantages. First, it has a complete description of the research work, reference materials, thoughts and ideas related to your work. Second, it provides a complete documentation of the work done and could be used as a source for writing your reports and other scientific publications. Third, lab notebook could be used as an evidence in the court for proving the intellectual property. According to U.S patent law states the inventor possessing is determined by the first to invent not the first to file. So, it's an important advantage of using a lab notebook and recording of your research work in it.

There are some advices to have a good lab notebook. The lab notebook should be well bounded that don't allow any pages to be ever removed. All pages should be numbered. The first page should contain your information; name, email, phone, etc. the second page should contain a table of contents, includes the title of each experiment and the page number on which the experiment begins. All entries should be written with a permanent marker and dated. All entries are noted down during the proceeding of the experimental work, not after the ending of the experiment. Feel free to record all your thoughts about the research work and to record everything you do. Don't worry about the mistakes; only record what you want in a readable way.

Second, examining the quality of the collected data

In order to find the right conclusion, you should use a good quality data. The reliability of the interpretations is mainly affected by both the validity of the assumptions and the quality of the data. There are different types of errors that can affect the quality of the experimental data. Scientists need to know these different types of errors to eliminate them and to get good data to draw the right conclusion.

Error in the data is the difference between the recorded measurements and the true value. Errors have two main types based on their source. The first type is the random errors which resulted from both the uncontrolled and unknown variables. The second type is the systematic errors which resulted from the measuring instruments. The researchers need to replicate their experiments to ensure the correctness of their measurements. The collected data from different replicates help in investigating if the experimentation process was done in the right manner or not. The differences between the measurements of the replicated experiments used to detect the degree of the precision and accuracy of the data.

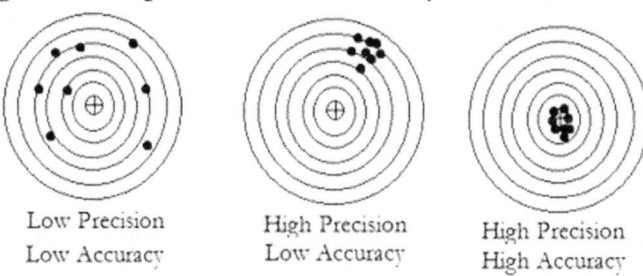

Low Precision   High Precision   High Precision
Low Accuracy   Low Accuracy   High Accuracy

⊕ **The true value**   • **The measured value**

Precision is the measure of the scatter, dispersion or replicatability of the replicated measurements. Accuracy deals with how close is the measured value to the true value. Data quality could be detected from its precision and accuracy degree. Random error could lead to both lower precision and accuracy. Systematic error mainly effect on the accuracy of data.

To get a good quality data you have to eliminate both the random and systematic errors. The random error is resulted from both the uncontrolled and unknown variables. This type of error couldn't be avoided, but it could be corrected by statistical analysis of the replicated measurements, the more the number of replications, the more the statistical corrections. In contrast, systematic errors couldn't be corrected statistically. This type of error arises from wrongs with experimental instruments such as; wrongs with the instrumental calibration or using the instrument in a wrong way. This type of error couldn't be corrected statistically because the resulted data are always offset the true value in the same direction. When the experiment replicated in the same way, the systematic errors will appear again. The only way to avoid the systematic error is by avoiding its causes.

Third, finding the correct conclusion through logical thinking

Scientific research, as you have known, consists of definite sequential steps. These steps begin with an interesting question about a particular phenomenon. Next step, you try to suggest an answer for this question and to formulate a testable and falsifiable hypothesis. Then, you should empirically test this hypothesis by testing the correctness of its deducted predictions. After that, you should record the resulted measurements of this empirical study, which are called the empirical data. Next, you need to examine the quality of these data based on the replication of these measurements. The final step is to draw the right scientific conclusion form these data. Data are recorded measurements that don't have meaning. To transform these data into meaningful information, it must be interpreted. The main goal of drawing a conclusion is to know if the hypothesis deducted predictions are correct or not. If the data agree with the predictions, it means that these data only support the formulated hypothesis, not confirmed it. If the resulted data disagree with the predictions, it means that these data prove the falseness of the hypothesis.

In The End

In the conclusion step your main goal is to draw a right scientific conclusion for the resulted empirical measurements. You have to interpret these data to get meaningful information by relating them to your formulated hypothesis. To know finally, the correctness of your suggested answer is supported or not.

6

Step Six:
Share Your Results

Sharing of knowledge with others is the fuel of the science progression. You not only have to share your findings, but you also have to choose the best way of sharing.

6.1. Knowledge Sharing....

"Gaining knowledge is the first step to wisdom. Sharing it, is the first step to humanity"

Unknown

Knowledge sharing is a human trait that began with the first moment of the human existence on the earth planet and it continues with the human life. As Aristotle said that "all humans by nature desire to know", also it is clear that they, by nature, want to share what they knew and what they found with others. There are nearly internal powers that push us to share our knowledge.

Importance of Knowledge Sharing

The processes of knowledge acquiring and sharing are very important for human life to the degree that the existence of humans cannot continue without them. So, we have them naturally.

For human life

Knowledge sharing began from the first moment of humans on the earth. Parents share their knowledge about life to their children to help them live a good life and to know how to deal with their universe and the nature around them. No one can learn everything needed in his life by himself. We need to know the history and the experiences of others to know how to live a good life. There must be knowledge sharing between us to learn how to live a good life and to avoid the mistakes which others had.

For progression of science

Also, knowledge sharing has a great influence on the progression of science and the scientific knowledge development. Scientific knowledge was developed previously by accident to some people, then they were sharing it with others. But now, it's the main goal of the scientists who are working to understand our universe and to solve its unknown secrets and to use these secrets in the benefits of mankind. Those scientists don't begin their work from zero. As I talked before that scientific knowledge is in a continuous progression and the development of this knowledge is based on the cooperation of efforts of different scientists. Every scientist tries to complete the work of the previous ones. They are completing the work of each other. If they didn't share their scientific knowledge with others, no one will be able to add new knowledge and they will always be at the zero point. So, to efficiently cooperate together, their scientific knowledge must be shared between them.

Knowledge and Knowledge Sharing

Knowledge is the experiences, information and skills that we have and knowledge sharing is the process of transferring this knowledge to others. The first step of sharing knowledge is to acquire knowledge itself after that you can transfer it with the others. We have acquired our knowledge from different sources such as our life experience through different events that we pass through and the most important method is by learning from others experiences. If no one shares his knowledge with others no one will gain knowledge. The scientific knowledge, that mankind have, was developed through different generations. This knowledge is not only passed through one generation to another, but it is also in refining process all the time. Books, for example, is an important way for knowledge sharing. The books contain a collection of different types of human experiences. These books were written by persons who decided to share their knowledge with others. These books contain parts about the author's experiences beside other knowledge related to other persons that also decide to share knowledge with others.

Progression of knowledge sharing

The progression in knowledge development mainly depends on the progression of knowledge sharing. This explains what we see in the current massive progression in the human knowledge development is due to the high progression in knowledge sharing. Progression in knowledge sharing was passed through different stages. It is based mainly on three factors and the development in them lead to the development or progression in knowledge sharing.

Recording methods - Writing Invention

As it was mentioned previously that human knowledge sharing began with humans from their existence moment on the earth planet. Knowledge sharing is in a continuous progression all the time. First, it was occurring orally from who have this knowledge to others and this type of knowledge occurred between parents or older brothers with the younger ones and also occurred till today and from the chief of the works with their new employers. This type of knowledge transfer stilled the only mean for knowledge transfer until the invention of writing. Writing invention is one of the most important achievements of mankind. The invention of writing wasn't occurred directly. It was passed through different stages and in different places of the world. It firstly began by pictures and signs and passed through different steps until the phonetic writing system, then the beginning of the alphabetic writing. After the writing invention, it became easy to note down the knowledge that anyone has. Writing became an effective method for knowledge sharing. It's a very good method to save and record the knowledge. Before writing invention, saving of knowledge was the task of an individual's brain that may be forgotten or lost by the dying of who has this knowledge. But writing invention helped in saving knowledge for different generations. After the writing invention the old nations began to record their knowledge and their history on different means to save it for the next generations. By theses writings, we have known the history of the different old civilizations. Writing was the first efficient method to record knowledge. Now, there are other ways and different facilities that help in recording and saving knowledge as audio, video, etc. Progression in the methods of recording knowledge has a great influence on knowledge sharing progression.

Copying Methods - Printing Press Invention

Another important step in the history of the knowledge sharing is the invention of the printing press. The printing press was invented by Johannes Gutenberg in 1440. It was one of the important factors behind the European scientific revolution. Knowledge began to spread everywhere quickly and accurately. Many numbers of books of a secular nature were printed. Books cost reduced. The copying rate of the books become in high amount and these increases the distribution of these books and the knowledge transfer become more easy

Transportation and Communication methods

Transportation and communication methods have a great influence on knowledge sharing. Advances in these methods decrease the time required to transfer knowledge between different nations. These transportation and communication methods passed through main milestones; the invention of **the steam engine**, the invention of **the electric motor**, using of **the wireless waves** in communication such as radio, television, etc. And the most important stage, which we are living in today, is the invention of **the internet**. Through internet knowledge can be transferred between different nations very easy and in no time.

In The End

We are living now in the golden age of knowledge sharing and this progression in knowledge sharing lead to high progression in science and knowledge development. Advances in recording, copying, transportation and communication methods are the main influential factors that influence the knowledge sharing progression that reach the level that we have today.

6.2. Scientific Communication....

"Whenever I found out anything remarkable, I have thought it my duty to put down my discovery on paper, so that all ingenious people might be informed thereof".

Antonie van Leeuwenhoek

It is essential nowadays for the researchers to spend a large part of their time communicating with the scientific community and scientific organizations. The continuity of their scientific researches mainly depends on this communication. Firstly, they need to apply for getting funds for their scientific work. Secondly, the researchers and scientists have a high level of experiences in their field and they communicate to deliver this knowledge and to share it with others. Thirdly, they need to share the findings of their research work. This is the most important goal of the scientific communication. It doesn't mean anything to find something or to discover something if you don't tell others. In E.H. Miller 1993 *"If it wasn't published, it wasn't done"*. The process of sharing knowledge is the fuel of the scientific progression. Science development is based on the cooperation between the scientists and this process of sharing knowledge helps all of them to know and continue the process of knowledge production and development. Sharing your findings not only helps others in their research work, but it also reserves your rights to be the first who find these results.

The process of scientific communication has varied methods. Each method differs according to its goal. Scientific publication is the method by which scientists report their research work and explain their findings. There are two main ways of the academic publishing; **Peer review journals and scientific conferences.**

Article in peer review journal

Peer review journal is the most common way for publishing the scientific works. The peer review journal is a periodically publication concerned with publishing a peer reviewed scientific articles in a particular area of science. The Peer review process is a review of the researcher's work by an expert in the same field. Its goal is to assess the quality of the scientific publications and it's called peer review because the review is done by the author's peers.

Published articles

The published scientific papers provide information in details about the different parts of the research work. It contains introductory part to explain the main problem of the research, the proposed hypothesis and the suggested plan of work. Then, there is the materials and methods section that include what you did in the research work and how you did it. Then, there is the results section in which you can present your findings with the appropriate manner such as tables, figures, charts, etc. After that, there is the discussion section in which you can analyse the results to find the final conclusion through relating these findings to the main problem and the proposed hypothesis. Research articles also include a reference section in which you mention the references that your work was based on it.

Write a research paper

The process of writing a research paper contains many details, but here I want to introduce its main characteristics only.

The Format: There are different formats for the articles of the peer review journals based on the journal that you will publish in. But, there is a general format for it that include;

title, abstract, introduction, materials and methods, results, discussion, conclusion and references

Peer review steps: When the article is submitted to a peer review journal, the editor sends it to the reviewers who are experts in this field to get their opinion about

- **The paper quality such as its accuracy, validity of the research methodology and procedures.**
- **Relevance to the journal field**
- **Appropriate for the journal specifications**

Journal ranking: The first peer review journal was Philosophical Transactions of The Royal Society in 1665. Now, there are large numbers of the journals in each field. Their ranking is based on a standard evaluation system. Every journal has an impact factor which is based on the average number of the citations to the articles published in this Journal. Citation of articles is the referring to this article in another scientific publication. Scientists also could be evaluated based on the citations of their published articles. This method is called H-index, which measures both the productivity and the citation impact of the published work of a scientist.

Conferences

Scientific conferences are another way to share your research findings with the scientific community. Scientific conferences are meetings of the researchers that have the same interest to share their work and to discuss together.

Present your work

In the conferences, there two ways to present your research work. The first way, is the talks, which are divided into two types. Invited keynote speakers' talks are 45 minutes long for each one with 15 minutes for discussion. The other type is the panel sessions which usually include 3-4 speakers, each one talks for 15-20 minutes with 2-5 minutes for discussion. The second way, is the scientific posters which are poster paper has information about your research work printed in 119 X 84 cm and attached to a stand in a specific hall. Posters act as an introduction to your new research work. They presented by a representative of the research team at a poster session to the conference attendance.

Conference advantages

Scientific conferences could provide great opportunities for you if you well prepared. They are really important opportunity to discuss many of the scientific topics and to share your results and your thoughts. At conferences, you can meet the key figures and the experts in your field. This type of communication is very important. It provides an immediate evaluation for your work and a feedback from the experts in your field. It acts as a preliminary step before a complete proceeding of your work. In addition, there are other advantages for the scientific conferences. It will help you to be aware with the recent developments in your interesting field. Also in conferences, you can build a network and contacts with the distinguished experts in your field who you can ask their advice in the future. So, you should be well prepared and put your plan based on the schedule of the conference.

Find a conference

There are many websites on the World Wide Web that offer the service of announcing upcoming scientific conferences and scientific events. They also categorize these events according to the field of specialization and the date and by country.

Rules for Conducting of The Scientific Research

There are different methods of the scientific communication and every method has its specific characteristics and formats. But, there are general rules for the conducting of the academic researches. According to **FEDERAL POLICY ON RESEARCH MISCONDUCT**, Research misconduct is defined as fabrication, falsification, or plagiarism in proposing, performing, or reviewing research, or in reporting research results.

- **Fabrication is making up data or results and recording or reporting them.**
- **Falsification is manipulating research materials, equipment, or processes, or changing or omitting data or results such that the research is not accurately represented in the research record.**
- **Plagiarism is the appropriation of another person's ideas, processes, results, or words without giving appropriate credit.**

In The End

Scientists' Communication with the scientific community is very important. There are different ways for the scientific communication. The scientists should choose their right way to share their research work and avoid the research misconduct.

6.3. Scientists and Public Community....

"In science one tries to tell people, in such a way as to be understood by everyone, something that no one ever knew before. But in poetry, it's the exact opposite."

Paul Dirac

Now, we are living in the age of science. There is no doubt that the current progression of science completely changed our life. The role of science is presented in everything around us. It touches all different disciplines of our life. So, it's very important for the public people, to be able to live in this age of science, to understand science and to know how it can help them. This is one of the most important roles of the researchers and scientists. They should communicate with public community and share their knowledge and explain what science is.

Importance of the communication with the public community

Communication between scientists and the public community is very important.

To use the scientific products correctly

First, it will help the public use scientific products correctly. Scientists' goal is to discover the secrets of the universe and to use this knowledge in solving the public problems and to change their life to be better. If the public community can't understand what the scientists do, there will be no use for what they do. There is an amount of knowledge that scientists should deliver to the public community. It's very important for normal people to understand science to the degree that can enable them to use its products and its applications, which affect everything in our life, in a right way. So, it's very important for scientists to explain these findings to the normal people who are not specialized in science to be able to use science in an appropriate way in their life.

Help the decision makers

Second, scientists should share their knowledge and explain it to the public not only to use its products correctly, but also to help the decision makers to know what science can provide and what're the amounts of benefits that they can get from science? Communication between scientists and the decision makers is very important to explain the facilities that science can provide and to help them in taking the right decisions

Increase the research funds

Third, sharing of the scientific knowledge is not only helpful for public community; it is also helpful for scientific progression. The scientists need money to work and when the normal people recognize the importance of science they will support the funding process of the scientific work. Normal people can motivate their governments to increase the financial support of the scientific researches when they realize the importance of science and what science can provide for them. Public people should realize the importance of what you do if you want to continue in your work.

Stimulate the scientific curiosity

Fourth, sharing some interesting findings and talking about the scientific concepts with the public community could stimulate their scientific curiosity and increase their interest to work in science. Their efforts could help in increasing the progression of science and the continuity of scientific research that is very essential for the development of any community.

Ways to Communicate with The Public Community

To effectively communicate with public community you have to know two important things. First, public people don't have your scientific background. Second, not all of them have the same degree of the scientific curiosity that scientists have. So, based on these two facts you should carefully choose what you can share and how. Then you should choose the way to stimulate their curiosity to know and how to make them interest about what you say.

You should choose carefully the level of details that could be delivered to the public people. Concentrate in your communication on the importance of science and what science can provide for them. You should clarify your ideas with well understood examples. It's very important to know that people don't understand well the indirect importance of science. So, try to express the direct and the clearest examples for the importance of science which are related their everyday life.

Public individuals are highly interested about scientists' news and stories more than their work. When Albert Einstein met Charlie Chaplin, Einstein said, *"What I admire most about your art, is its universality. You do not say a word, and yet, the world understands you." "It's true",* replied Chaplin, *"But your fame is even greater: The world admires you, when nobody understands you."* Nearly, all humans feel interest to know the personal stories. So, you can express and explain what you want to say by telling the stories and the different remarkable situations about the journey of research and scientific discovery.

You can communicate with public community through different methods such as; books for public or especially for young individuals, articles in public magazines, public lectures, through programs in radio and television, scientific documentary films and social media and internet.

In The End

Communication between scientists and the public individuals is not less important than communicating with scientists. You should communicate with the public community in a simple, clear, polite and interesting way.

6.4. Truth Shock....

"New ideas are not completely easy to accept, sometimes even by the brightest and most open of people."

J. Steinberger

The way of science has many challenges and the most important challenge appears after the finding of what you think it's the true answer for the problem of interest. Truth is very valuable. It's the most priceless thing that anyone can have. Finding the truth and searching for it, is the best life's activities, but it's not the whole story. The story has an important part that always begins after the finding of this truth and the beginning of delivering it to others.

Rejection and The Great Scientists

I want to ask you a question. What do you expect about the response that you will face when you find a great idea? Do you think that you will be celebrated for your new idea? I am sorry to tell you that many of the great discoveries that changed our life, was highly rejected in the beginning. **You have to know that the great scientists not only find great discoveries, but they also face higher degrees of rejections and attacks due to their new ideas.** Many of the current highly respected scientific discoveries were highly rejected at the beginning such as rejection of Copernicus and Galileo for the idea that the earth revolves around the earth, Louis Pasteur for the germ theory of diseases, Boltzmann for the idea that the matter is comprised of atoms and molecules, Edward-Jenner for the vaccination, Hans Krebs for the Krebs cycle, Julius Myer for

Law of conversation energy and so many other examples. These are only a few examples for the ideas that were highly rejected in the beginning. So, don't be surprised and prepare yourself for that response. These attacks and rejections came from either the scientific or public community. Truth needs a brave person who can defend for it. You have to search for the truth and when you find it, you should defend for it.

The Goal of The Rejection

The community rejection is a very painful situation. Humans are social individuals. They need the acceptance of the community. This need for the community acceptance has advantages. It makes everyone avoid doing bad things in order to not lose this social acceptance. So, it acts as a defense mechanism for the community from wrong activities and wrong ideas. But, if this need for social acceptance will make you change your opinion about what you think it's the right truth and accept the community wrong idea, here is the most dangerous disadvantage. The main goal of the rejection is to make you under a hard pressure, to change your thoughts and to accept the current thoughts even if you are not convinced with them. You may face many rejections in the way of science begins with your refusing to walk through the normal people way. Then, when you find new concepts or discover new thing that is against the current scientific paradigm. So, be prepared for all types of rejection.

The rejection has different forms. Some scientists faced the criticism or ignorance or they were excluded from the community. Some were accused with craziness, some were accused that they are against the religion as a non-believers. Some forms exceed that to more violent actions such as body harming and so on. All this forms had occurred for many scientists.

Reasons for Rejection

There are three main reasons for the rejection of the great ideas.

Novelty

This is the first reason and the most important one behind the rejection. Uncertainty always related to the novelty. A study by Pennsylvania university demonstrated that people always feel uncertain about the novel idea. People always reject everything new different from what they knew. People don't trust in anything new, it's a human trait.

Threat others' benefits

Another reason behind the rejection of the new idea is the fears of the loss of benefits. Truth may threat others' benefits. So, they may claim that the idea are not right or not useful. Also, the public opposition of a new idea may be motivated by men who fear to lose their benefits

Misunderstanding

Misunderstanding is the common reason for the rejection of the new ideas, especially the rejections that come from the public community. This misunderstanding arises due to the difference in the ways of thinking between scientists and public people. Not all the public people think in a scientific way. Normal people don't have the scientific background that can help them to understand easily the scientific ideas. In many cases, scientists fail to convince people with their idea. They find difficulties in speaking with them about the details and give them the end result or the end conclusion. Some scientific findings may represent a shock for public community because it may be against what they thought or against their wrong thoughts. These ideas will be faced with high resistance and rejection.

Differentiate between Two Types of Rejections

You should differentiate between two types of rejections. First, the rejection from a person who is convinced with another idea and defends for it because he thinks it's the right one. Second, the rejection from a person who fears loss of his benefits of the current established idea. You can discuss with the first type what support your idea and what contradict other idea. But, the second type will not accept your idea. Logical thinking isn't useful to him because he defends for his benefits not for the idea. You can differentiate between these two types based on their way of rejection. The first type will discuss the idea. He would talk about what support his idea and what contradict your idea. The second type will concentrate on the personal attack because he doesn't concern with the correctness or falseness of the idea itself. He mainly wants to save his benefits.

Rejection is The Beginning of Success

You must remember that rejection is not bad. Most of the great people who changed the world were highly rejected. Accept the rejection; it's normal to be rejected due to your difference. Rejection is a good thing. Rejection is your way of success. Success way always begins by rejection. First, you reject an established wrong situation. Then the others will reject you because you don't want to walk with them and to follow their wrong way. But in the end, they will follow you. Arthur Schopenhauer said that "All truth passes through three stages. First, it is ridiculed. Second, it is violently opposed. Third, it is acceptable as being self-evident."

How to Deal with The Rejection

Imagine that you are on a ship in the sea and you have an apparatus that have the ability to detect the presence of the large bodies in the way of the ship. And you find that there is a mountain in the way of the ship and the current high speed will made the ship collide with this mountain. You call to change the direction of the ship and nobody else agrees with you. What will you do in this situation? Will you resist this rejection or you will agree with them and accept crashing of the ship with the mountain? You have to know that, soon or later, the truth will appear one day and everything will be clear. So, you must be with the truth side and don't scarify with the it for any other thing, whatever it was. The great scientists had the courage of facing the rejection. They didn't fear it. It was begun with them from their early stages of life when they didn't fear to express their different views, thoughts or to be different from the other people.

Look at The Rejection

Don't believe their opinion about yourself. Your value is not determined by the opinion of others. Take care; don't change your opinion about yourself due to their opinion. The real situation is only different positions; you see the idea differently from them. You should understand correctly the process of rejection. Don't deal with rejection as it's an attack against you. No, it's an action against an idea. Feel easy, you don't matter to them, they only want to face this idea, they only reject the idea. If they increase the dose of the personal attack, it's a sign that your idea is very strong that they cannot argue against it. But take care, others could believe their claims.

In The End

You have to know if you decide to be accepted by leaving what you think it's the truth and agree with their wrong thoughts you will face a harder pain than the pain of the others' rejection. This will be the rejection of yourself. It's very harder than any other type of rejection that you may face. The scientist's word isn't like any other word, It has a great effect on the scientific and public community. You must watch your declarations and take your responsibilities.

7

Advices for The Way

There are some advices that can help you to complete your way and reach your goals.

1) Your Purpose....

To succeed in the way of science and to live a happy life, you have to find the meaning of your life. Why you are alive. What's the most valuable thing that deserves to spend your life for? You must answer this important question. Its answer is your general purpose of life. It must represent a true value for you and agree with your beliefs about the real mission of humans on the earth. Then, you should put your passion for science in the service of this purpose. Your purpose has an important effect on your success in the way of science. **Why you do, is more important than what you do.** Why you go through this way. What do you want to achieve? And Why? If you decide to take the way of science you have to know it's just a way. And every way is taken in order to reach a specific destination. Your destination is the specific goals that you want to achieve. To succeed and to be satisfied with your life, your goals must agree with your general purpose of life and you should have companions who share the same life meaning.

2) Persistence….

During your movement in your way toward your goals, you may face many challenges. You may have a plan of action, but you have to know that it doesn't always work as you expect. Sometimes, you may find yourself doubting, will you achieve your goals or not? You may ask yourself, am I doing the right actions? And if I am doing the right actions, why it doesn't go well. All these thoughts appear when you don't find any results of what you do or the results aren't as your expectations. At this moment, what you should do is to keep moving toward your goals and never ever quit. Do you know the Chinese bamboo tree? It's a very distinguishing plant. Planting of this tree needs a continuous irrigation and fertilization for its seeds for nearly five years without appearing of any results above the soil. So, who decides to plant this tree will wait for five years without finding any substantial results of what he are doing. Could you imagine that? Then, after the passing of these 5 years of hard working the sprout begins to appear and it takes only 6 weeks to reach 90 feet long. It's amazing. But, do you think that it needs only 6 weeks to reach 90 feet long. No, it's not true. It needs 5 years and 6 weeks to reach 90 feet long. The work begins from the first moment after the planting of the seeds. But, you can't see the result of this work because it is under the soil. It's also the same for your dreams. The more you wait and the more effort and time you spend, the more you will get. It's a very difficult situation when you don't get any results for what you do. Humans always want to get quick fructification for what they do, although it's not the reality. Actually, it takes time to achieve your goals. No success without persistence. You must trust in God that someday you will be rewarded for all what you do. You will see the results for everything you do if you persist. But if you give up, you will waste your previous works. The only case for quite, is when you find that there is no other available step that you can take. But, **if there is another step, take it. You may be very close.**

3) Perspiration....

It's very important to know that nothing in this life for free and there's no success without hard working. Some people thought that scientists are very relaxed people, all what they do is to think and by thinking only they find solutions for the scientific problems. It's completely wrong. Scientific research is a continuous process of working and working until the finding of the answer. It needs great efforts and time. You may spend weeks, months or in some cases years – working day and night, studying, searching, thinking, doing experiments, writing and repeating that more than one time – searching for the answer of only one question. It's really very hard work, but it's also very enjoyable for who has a passion for it.

4) Questions Never Stop....

In the end, you have to know that the process of the scientific research never ends. All the time, there is continuous arising of new questions. When you find a solution of a scientific problem, this solution will stimulate arising of other questions that needed to be answered and the new answers will stimulate arising of other ones and so on. There is no end to the questioning process. The knowledge that we have about the surrounding universe is very limited. If it compared to our ignorance, it will not represent anything. You should always remember this fact on your way of science. These feelings of ignorance are the motivational factor for the continuity of the scientific research and its very important to still enthusiastic to be always on the way of science.

Bibliography

*Bryan Bunch and Alexander Hellemans (2004) **"The History of Science and Technology"** United States of America by Scientific Publishing, Inc.*

*Charles R. Gibson (1921) **"Stories of Great Scientists"** London by Seely, Service Co. Limited.*

*Claus Ascheron and Angela Kickuth (2005) **"Make Your Mark in Science"** Hoboken, New Jersey by John Wiley & Sons, Inc.*

*Craig Loehle (2010) **"Becoming a Successful Scientist: Strategic Thinking for Scientific Discovery"** New York, United States of America by Cambridge University Press.*

*Francis J. Rowbotham (1918) **"Story-lives of Great Scientists"** London, Wells Gardner, Darton & Co. LTD*

*Geoffrey Marczyk, David DeMatteo and David Festinger (2005) **"Essentials of Research Design and Methodology"** Hoboken, New Jersey by John Wiley & Sons, Inc.*

*Gregory N. Derry, (1999) **"What Science Is and How It Works"** Princeton, New Jersey by Princeton University Press.*

*John Adair (2007) **"The Art of Creative Thinking"** Great Britain and the United States by Kogan Page Limited.*

*John P. Dickinson, (1986) **"Science and scientific researchers in modern society"** France by Division of science and technology policies, UNESCO.*

*John Waller (2002) **"Fabulous science: Fact and fiction in the history of scientific discovery"** New York, United States by Oxford University Press Inc.*

Bibliography

*Juan Miguel (2009) **"Rejecting and resisting Nobel class discoveries: accounts by Nobel Laureates Scientometrics"** Campanario Vol. 81, No. 2 549–565*

*Keith J. Holyoak and Robert G. Morrison (2005) **"The Cambridge Handbook of Thinking and Reasoning"** Cambridge, UK: by Cambridge University Press.*

*Martin F. Rosenuman (2001) **"Serendipity and scientific discovery"** creativity and leadership in the 21st Century Firm, Volume 13, Pages 187-193 Elsevier Science Ltd.*

*Sabry Eldmrdash (2008) **"Scientific anecdote: Intro for Teaching Science"** seventh edition Cairo, Egypt. By Dar Elmaaref*

Web sites:

***"Understanding Science."** (2015). University of California Museum of Paleontology, retrieved from http://www.understandingscience.org*

*Anne Egger, Ph.D., Anthony Carpi, Ph.D.. (2011) **"The How and Why of Scientific Meetings"** Visionlearning Vol. POS-3 (3) retrieved from http://www.visionlearning.com/en/library/Process-of-Science/49/The-How-and-Why-of-Scientific-Meetings/186*

*G. Carboni, (2006), **"The History of Writing"** Translation edited by Karyn Loscocco (2008) retrieved from http://www.funsci.com/fun3_en/writing/writing.htm*

*Jessica Olien(2013). **"Inside the Box: People don't actually like creativity"** retrieved from http://www.slate.com/articles/health_and_science/science/2013/12/creativity_is_rejected_teachers_and_bosses_don_t_value_out_of_the_box_thinking.html*

*Martyn Shuttleworth (2009) **"History of the Scientific Method."** retrieved from https://explorable.com/history-of-the-scientific-method*

*Salman Khan (2014) **"The Learning Myth: Why I'll Never Tell My Son He's Smart"** http://www.huffingtonpost.com/salman-khan/the-learning-myth-why-ill_b_5691681.html*

The Author

Sherif Elkaffas is a Molecular Microbiologist. He was born in Disuq City in Egypt. He is interested in science and scientific research. He studied Microbiology in Alexandria University. He received the Bachelor of Science in Microbiology in 2008 and the Master Degree of Science in Microbiology in 2014. He worked in the Genetic Engineering and Biotechnology Research Institute of the SRTA City in Alexandria from 2009 to 2014. He assisted in different scientific research projects. He has got a good experience in the scientific research. He has wide readings about the history of science, the distinguished scientist and the scientific research process. He is mainly concerned with finding the rules of the science game and understanding how science work.

For contact with the author:

Email: Sherif.mostafa87@yahoo.com

www.ingramcontent.com/pod-product-compliance
Lightning Source LLC
Chambersburg PA
CBHW070810180526
45168CB00002B/557